THE TSAVO STORY

THE
TSAVO
STORY

Daphne Sheldrick

COLLINS AND HARVILL PRESS
LONDON 1973

ISBN 0 00 262801 5

Set in Monotype Baskerville

Made and Printed in Great Britain by
William Collins Sons & Co Ltd, Glasgow,
for Collins, St James's Place and
Harvill Press, 30A Pavilion Road
London SW1

For all who have had a hand in the making of Tsavo, and particularly for the elephants, whose contribution has probably been the most significant.

To see a World in a Grain of Sand,
And a Heaven in a Wild Flower,
Hold Infinity in the Palm of your Hand,
And Eternity in an Hour.

Auguries of Innocence
WILLIAM BLAKE

Contents

Illustrations

Acknowledgements

I would like to record my sincere thanks to all who have contributed towards this book; to Bill Travers and Simon Trevor who persuaded me to write it in the first place; to Gerard Vienne, Marion Kaplan, Bill Woodley, Bernard Kunicki, David Paynter of the Argus Africa News Service, my husband, and again Simon Trevor for the use of photographs; to my mother, Mrs Bryan Jenkins, for her painstaking sketches made even more a labour of love by failing eyesight and to my father, whose suggestions and criticisms inspired many ideas. But I probably owe most of all to the animals that have enriched our lives over the years, and of course to David, whose brains I have largely picked, and who has made this story possible.

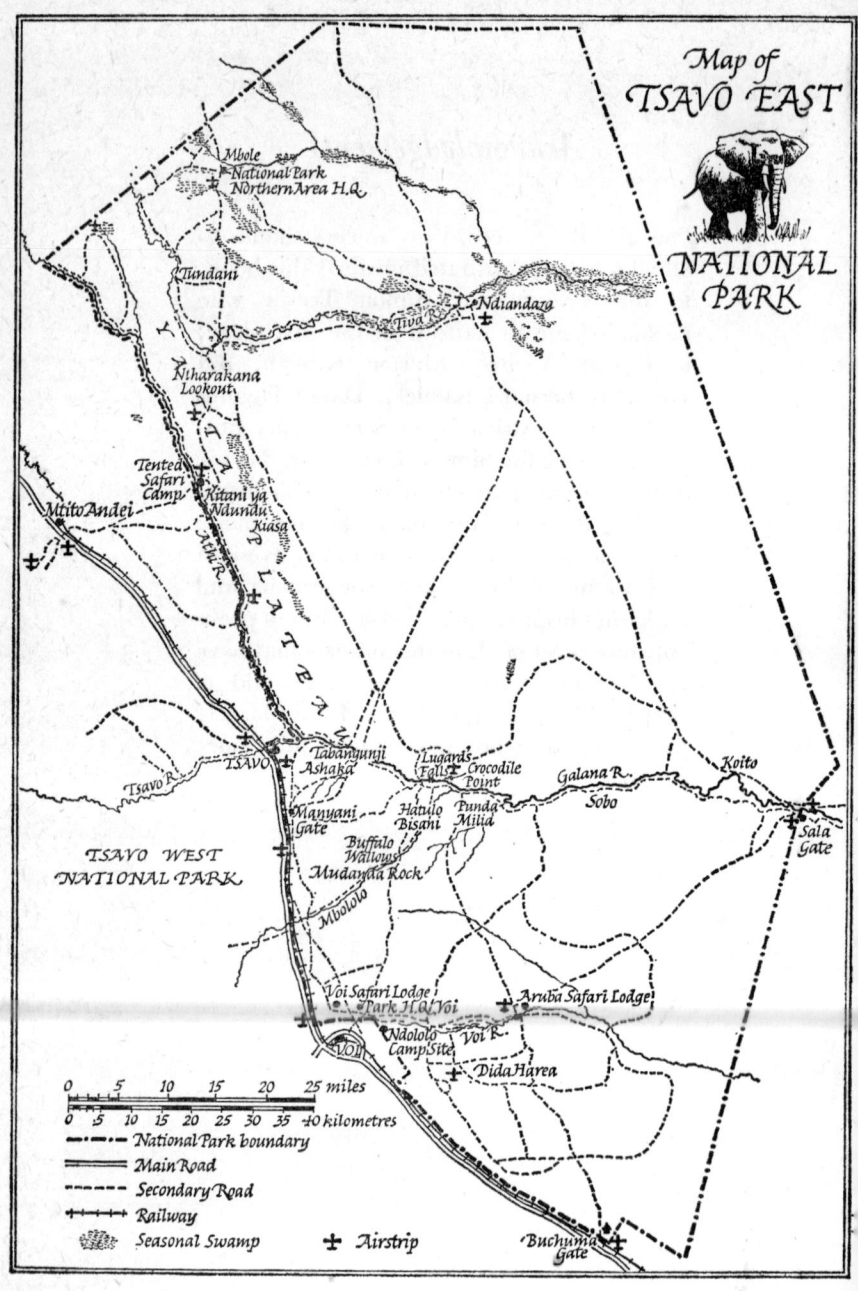

Map of
TSAVO EAST

NATIONAL PARK

Mbole
National Park
Northern Area H.Q.

Tundani

Ndiandaza
Tiva R.

Ntharakana
Lookout

Tented
Safari
Camp
Mtito Andei
Kitani ya
Ndundu
Kiasa

Tabangunji
Ashaka
Lugaris
Falls
Crocodile
Point
Galana R.
Koito
Tsavo R.
TSAVO
Sobo
Manyani
Gate
Hatulo
Bisani
Punda
Milia
Sala
Gate
TSAVO WEST
NATIONAL PARK
Buffalo
Wallows
Mudanda Rock
Mbololo

Voi Safari Lodge
Park H.Q. Voi
Aruba Safari Lodge
Ndololo
Camp Site
Voi R.
VOI
Dida Harea

0   5   10   15   20   25 miles
0 5 10 15 20 25 30 35 40 kilometres
Buchuma
Gate
National Park boundary
Main Road
Secondary Road
Railway
Seasonal Swamp   Airstrip

Introduction

AT the turn of this century, early travellers to East Africa recorded with incredulity the wealth of wildlife that roamed the great plains in tens of thousands. They spoke of the 'extra-ordinary and almost incredible quantity' (Sir Charles Eliot); of areas 'literally crawling' with game. It was as though the spectacle left them bereft of words that could adequately describe what they saw. Sir Frederick Jackson wrote in 1906 that 'game was found throughout three-fifths of the whole Protectorate in very great variety and more or less plenty, and in about one fifth in great quantities'.

My great-grandparents were amongst those fortunate to have been in time to witness this wonder. They arrived in the Colony in 1907, complete with their wagons, their stock, their horses, and their 'know-how', eager to build a country and establish for themselves a new life. The land they were first allotted by the Government lay in the heart of the Masai country, even now one of the finest faunal areas in East Africa, so it is not difficult to picture what it must have been like in those days. It must have looked like paradise. Softly, rolling plains rippled gently as the grass, which resembled fields of ripe wheat, moved with each breeze. Tall, flat-topped acacia trees, the essence of Africa, occurred in shady stands on the edges of the plains. Thicketed *luggas* divided one plain from the other, and undulating slopes of low hills were lightly wooded. Those fertile plains pulsated with an abundance of life; enormous herds of zebra, wildebeest, eland and kongoni opened up to allow the column of new settlers passage, closing again behind them like a living curtain. Delicate little Thomson's and Grant's gazelle jinked and stotted across the plains in their thousands; huge mobs of ostrich ran alongside the wagons, beautiful impala leapt with perfect grace and sheer exuberance, while plum coloured topi stood sentinel on low anthills. Great herds of buffalo dotted the edges of the plains, retreating into the cover of the thickets as the wagons slowly wound their way towards their destination.

My forefathers must have gazed upon this sight with mixed feelings, for although it certainly gave an illusion of a land of plenty, the problems were equally plentiful and also in evidence. Lions prowled those plains, not in tens, but in prides of twenty and more. Furtive hyaenas howled all night in the wake of the wagons, and the dark silence was shattered frequently by the blood-chilling, sinister cough of a leopard at close quarters, answered by the nervous, restless stamping of the horses and cattle. Savage, spear-bearing Masai tribesmen, plastered in red ochre and mud, collected in groups to stare and chatter wherever the column halted, viewing the white man's incursion into their country with curiosity and a hint of hostility.

The nearest doctor was a week's wagon journey away, over a rough and dusty track, so that one had to be able to survive this ordeal to get a serious injury treated. There were no shops where essentials could be purchased, no schools where the children could be tutored, and no near neighbours to call on for help. The cattle, infested by ticks, succumbed to mysterious diseases and died, as did the sheep. In fact, Masailand turned into a settlers' nightmare. My great-grandfather therefore welcomed the unexpected news that the Government had decided to move him to an area known as Laikipia, so that his land could be handed over to the Laikipia Masai, who had found themselves isolated from the main tribe when the intervening country had been thrown open to white settlement. So the entire family, and what remained of their livestock, wearily retraced their steps to put down their roots in Laikipia.

Those initial difficult years in Masailand must have appeared like a bad dream to my great-grandparents, but they provided a fund of fascinating memories that were passed down through each generation; memories of those great herds of antelope, of hair-raising encounters with dangerous animals, of having to walk on foot many miles at night after one's horse had fallen prey to lions; of occasions when one even found oneself galloping amongst stampeding zebra with the uncomfortable feeling that your horse would be seized by a lion at any moment; of being tossed by buffalo and treed by rhino, lost and benighted in the bush, and stuck in swollen rivers.

Then came two world wars, and in the great herds lay an endless larder from which cheap protein for prisoners and troops could be drawn. Game was looked upon as vermin in the European settled areas; a threat to a farmer's stock by harbouring disease, and a threat to his land by competing for grazing. As always, a surfeit of anything tends to be taken for granted. Life was cheap, the stocks of game seemed endless; there for the taking; a source of free meat and free sport for anyone who could hold a gun.

And so, slowly those great herds began to dwindle, eroded by the impact of civilization, and with each year that passed, the numbers grew fewer, until people suddenly wondered in astonishment where all the animals had gone! The remaining herds now contained only a shadow of their former numbers, and great stretches of land lay empty and still. Only when small pockets of game struggled for survival did people begin to think in terms of conservation, becoming conscious of the need for protective legislation to ensure that this natural heritage would not disappear forever. There followed a subtle and successful attempt to rouse public opinion by Colonel Mervyn Cowie, through a series of provocative articles to the Press, in which he suggested that game should be exterminated to make way for progress. This had the desired effect, and the outcry which answered this suggestion forced a reluctant Colonial Government to act, more to appease public opinion than to preserve the game. A Game Policy Committee was appointed charged with the task of investigating the possibility of establishing National Parks in 'suitable' areas. The 'suitable' areas, however, were not those that supported the largest remaining concentrations of wildlife, or the greatest variety of animals, as one would suppose, but rather those that were unlikely to be required for any other purpose. That part of the country most fitting for a wildlife sanctuary was under treaty to the Masai people, who were not prepared to relinquish any part of it. Other areas that still harboured fair-sized wild herds, were well watered and fertile, and so earmarked for agricultural exploitation; yet other more marginal areas were set aside as possible ranchland; until, finally, there remained just one possibility – a large tract of arid land which was largely unexplored and uninhabited, lying between the Wakamba Reserve, the Masai country, the coastal strip and the Tanganyika border. This area, some 8,069 square miles in extent, of tsetse fly infested, desolate bushcountry, was unsuitable for domestic stock due to the prevalence of the lethal trypanosomiasis transmitted by the tsetse. It also had a very low and erratic rainfall, practically no sources of permanent water, and only very few inhabitants, most of whom comprised gangs of poachers, who roamed the country with bows and arrows in search of ivory and rhino horn.

Several early explorers had, of course, briefly passed through this area on their way inland, but none had covered it in any detail apart from one or two professional hunters during a quest for the large tuskers reputed to roam there. Apart from this not a great deal was known of what is now Tsavo. One man, G. G. MacArthur, a Senior Assistant Warden in the Game Department, perhaps knew the area better than anyone else, having travelled there during the war years when he wandered on foot through the flat interminable bush, pausing at the occasional muddy waterhole to slake his thirst. On these expeditions, he had come across some of those enormous elephant, and had experience too of the rangy lion, prone to man-eating, whose forerunners had brought the construction of the railway to a standstill at the turn of the century. He knew the fascination of this dry country with its red baked soil and twisted, termite castles, stark leafless trees and rugged rocky outcrops. The area was very rich in birdlife and supported a large population of elephant and black rhinoceros, but all other species only in small numbers. An occasional lesser kudu would bound silently away as he passed; sometimes a small group of oryx would dissolve into the bush, perhaps a graceful gerenuk pause long enough to be briefly glimpsed, or the odd buffalo retreat into a thicket. But there were no great herds in Tsavo, as had been the case elsewhere, and in fact, if one had been told to select an area in Kenya for its faunal attributes, Tsavo would, in all likelihood, have been rated amongst the last. But, as things turned out, it was the only 'suitable' area the Game Policy Committee could recommend for National Park status, and when the Government was satisfied that it was unlikely to be required for any other purpose in the foreseeable future, the necessary legislation was enacted which brought the Park into being in 1948; a faunal sanctuary not because of its wealth of wildlife, nor due to natural beauty of a scenic nature, but simply because this was the one place that could be spared for the remnants of those great herds, the only block of empty land that had no place in man's planning.

MacArthur, because of his knowledge of the area, had been consulted by the Game Policy Committee, and although he loved the remote *nyika*, he recognized that it was very much second best in a country with such a rich natural heritage. But he also realized that it was this, or nothing, and he lent strong support to the suggestion.

And so, Tsavo was born.

CHAPTER I

The Beginning

AFTER the war, MacArthur had retired from the Game Department and had established a professional hunting firm, known as Safariland. He had met David Sheldrick, who had just been demobilized, at the Game Department Offices, and had taken an instant liking to him, sensing in him the same intense love of nature and longing for wild places as he himself possessed. David was twenty-eight at the time. He had left the army as a Major, and was eager to work with wildlife. MacArthur took him aside and suggested that he join Safariland as a hunter, for a period at least, until he found something more to his liking. David was dubious about this proposition, and not prepared to commit himself at this stage, so the matter was left in abeyance while he went to London with the East African Contingent to take part in the Victory Parade. While there, however, a decision was more or less forced on him by way of a telegram from MacArthur, informing him that he had been booked for the Aga Khan's forthcoming safari. And so, David found himself a White Hunter.

It wasn't long before he became one of Safariland's top hunters, and MacArthur came to look upon him almost as a son. It was also not long before the constant, senseless slaughter palled, leaving David restless and discontented with a profession that offered no outlet for any creative ability, but seemed geared only to destroy the very thing he loved most; reduce to lifeless carcases creatures that were beautiful, wild and free, so that some opulent client could claim a pair of horns half an inch longer than those of his neighbour. MacArthur, a conservationist at heart, sympathized with the struggle that raged within David.

In due course, a Director of National Parks, in the person of Colonel Mervyn Cowie, was appointed, and applications for other vacancies were invited. Ken Beaton was selected to fill the post of Chief Warden for the Tsavo National Park, while Bill Woodley and Peter Jenkins found themselves the successful candidates for two vacancies as Junior Assistant Warden.

Meanwhile, David was engaged on a foot safari in the depths of Tanganyika, and was unaware that positions in the National Parks service had been advertised. To his immense disappointment, he arrived back only to find that the closing date for all applicants had already passed. He had been too late. It was MacArthur who went to the Director of National Parks and persuaded him to consider a late application in view of the extenuating circumstances of David's case, and who came back armed with the necessary forms for David to fill in. And so, thanks to him, David was allowed the opportunity to apply, and to his surprise, was selected and became one of two Assistant Wardens posted to the Tsavo Park to work directly under Ken Beaton, the Chief Warden.

By this time, Ken Beaton had undertaken a brief recce of what was now the Tsavo National Park, and it was very apparent to him from this trip that the task of developing it as a tourist attraction, was immense. He therefore recommended to the Director that it be divided into two Sections for administrative purposes; Tsavo West, comprising 3,000 square miles lying West of the main Mombasa/Nairobi railway line, and Tsavo East, embracing some 5,000 square miles of country East of the line, and that the two Assistant Wardens be upgraded to full Wardens, and one made responsible for each of these two areas.

The potential in Tsavo West was very obvious, for although the game was sparse, the area had tremendous scenic beauty with the softly undulating Chuyulu Range in the West, majestic Kilimanjaro brooding on the border, conical pyramids of bare lava ash rising suddenly from a dead flat plain, crystal clear springs bubbling from black, lava boulders, rugged *kopjes* of delicately tinted rock, and the palm fringed Tsavo River meandering leisurely towards the East.

Tsavo East, on the other hand, was the complete opposite, and here lay the greatest challenge. Not only was the game scarce and shy, having been harried by generations of poachers, but the area was almost totally featureless and devoid of scenic beauty. Only the long, flat Yatta Plateau relieved the monotony of a low sea of scrub, and only two permanent rivers slaked the thirst of this hot, dry land – the Athi River and the Tsavo River, which merged in the Park to form one, and at this point became the Galana. Fairly shallow in most places, the Galana River loitered lazily through the heart of Tsavo East all year round, and finally spilled into the sea near Malindi. There were other watercourses in the area, but these were seasonal, and flowed as muddy torrents for short periods during the rains only. Two of these could qualify as rivers; the Tiva in the north, and the Voi in the south; and although they were dry for the main portion of the year, the water lay just below the surface of the riverbed and could be utilized by game when it seeped into the holes excavated in the sand by elephants. Apart from this, no one knew very much about the area, for it had always been considered rather hostile, waterless scrub, not worth the trouble and risk of investigating; and it was difficult to visualize it as ever being a tourist attraction.

Ken Beaton considered the men at his disposal, and he selected David with Bill Woodley as his assistant, for the task of developing Tsavo East, while Major 'Tabs' Taberer and Peter Jenkins were posted to Tsavo West. And in his report to the Director, he wrote – 'They do not look upon their work as a means of earning a livelihood, but as a job in which they will take immense pride in achievement. The task before them is difficult and has many hazards, but I am convinced that they will do all in their power to make the Park a success.'

It was to take them twenty-two years of dedicated endeavour to surmount the obstacles in order to begin to achieve this aim.

The announcement that Tsavo East had been allocated him, filled David with immense disappointment. Although he hardly knew the area, he knew enough about it to appreciate what he was up against. Bill Woodley had already undertaken a foot safari into the remote northern part of the Park, and had brought back alarming reports of the largescale illegal poaching being carried out by Wakamba tribesmen, whose reserve bordered the northern Park boundary. David sought the advice of MacArthur, spending long hours with him that stretched well into the night; drawing out of him knowledge he had accumulated over the years – vital information concerning the waterholes, the game trails, the poachers and how they operated, and the animals in the area. What David learnt from these discussions didn't make him feel any happier, but merely confirmed what he had feared. However, with a lorry and six labourers at their disposal, he and Bill Woodley set out to establish a headquarters and develop Tsavo East, not really knowing how to begin.

Arriving at Voi, they called first on the District Commissioner, who reflected the attitude of a lot of Colonial Administrators and who did not try to hide the fact that he viewed the whole exercise with some cynicism, for what was the purpose of devoting a lot of effort, time, and money to trying to develop a vast tract of useless scrub so that the rear end of an occasional elephant could be seen as it disappeared into the bush? And who in their right senses would pay to go into such a place, anyway; to drive through miles of monotonous flat bush in temperatures upwards of 90°F, and probably not even see that? The District Commissioner, although very helpful, regarded David and Bill somewhat pityingly as they heatedly defended the cause of wildlife, and it was plain that he thought that the world was full of cranks, and Kenya had more than its fair share of them! David and Bill couldn't help feeling rather depressed as they walked out of his office.

They established a camp near Voi township where they spent that night. Voi boasted a handful of Asian-owned *dukas*, or shops, bordering a dusty track, whose shelves stocked a jumbled assortment of commodities ranging from sweets to hurricane lanterns and cement. There was a railway station; a small tin-roofed shed which served as a post office; a police station; various Government offices and a hospital under construction; a neat orderly War Cemetery, where an interesting assortment of people lay buried, including a holder of the Victoria Cross, and all that remained of a man devoured by a lion: his boots. Further along, the new Voi Hotel was also under construction, which had been designed on the back of an empty cigarette box, and looked as though it had, as well! Over the township brooded the twin massifs of Ndara and Sagalla, part of the Teita complex of hills.

The next morning, David and Bill followed a rough track which took them to where they thought the Park boundary must lie. Here they left the vehicle to seek a better vantage point on the top of Mazinga Hill in order to view their surroundings. The heat was already intense, even at that early hour, and by the time they had pushed their way through the thick undergrowth and thorn-barbed bush, scrambled over boulders and fallen trees, and made their way to the top of the hill, the perspiration was running in grimy rivulets down their faces. There before them lay their domain; a shimmering flat grey sea that took on a mauvish hue in the distance, and then turned to deep blue as it dipped over the line of the horizon. The path of the Voi River lay in sharp relief; a narrow belt of bright green that wound its way through the bush; and just visible on the northern horizon, lay the long Yatta Plateau.

Any enthusiasm the men had managed to summon since their arrival, seemed to drain from their faces, as they surveyed their Park for the first time. So devoid of features was the flat land before them that it could almost have been a scene from Mars. There were just miles and more miles of nothing: no hills, no plains, no lakes, and no roads. The immensity of what lay before them was enough to daunt even the stoutest heart, and David, for one, felt a strong inclination to give up there and then as his thoughts turned to Diana, his wife of one year, whom he had married during his hunting days and had not seen very much of during that time. She had welcomed the break from hunting and the thought of a more settled existence, but David was doubtful as to whether she would like this any better. But, the challenge presented by the scene was tempting, and he couldn't resist it. With military precision he faced it. They had come to establish a headquarters, and so far they had not made much progress in this direction.

Deep in thought, they retraced their steps, and proceeded to a Waliangulu settlement they had noticed the day before on the far bank of the Voi River. As they approached it, David studied a map.

'No doubt about it! This lot are squatting illegally on Crown Land, and are well within the boundary of the Park, too,' he muttered. He made a mental note to speak to the District Commissioner about the matter and to take steps to get this illegal settlement removed from the Park forthwith.

At the Waliangulu village, known as Ndololo, he and Bill questioned some of the men about available sources of permanent water within striking distance of Voi. News of their arrival in the district had travelled fast on the 'bush telephone', and dozens of curious faces peered with interest at the new 'Bwana Game', who posed a threat to their peace of mind, for most of the able-bodied men in the settlement were active poachers with many elephant on their conscience. However, they were eager to please, and volunteered to lead David and Bill to a spring they knew of called Pundamilia, where, they said, was a very good supply of surface water. The spokesman assured them that the spring would be ideal for their purpose, for the water was plentiful, clear and sweet. There were nice green trees too – and lots of shade – in fact, *mzuri sana*, Swahili for 'very nice'. Considerably cheered by the account of this unexpected oasis in the midst of the dry bush, David and Bill set forth to investigate, taking with them the Waliangulu spokesman as a guide. They drove along a narrow elephant path for some miles until the bush became so dense that no further progress was possible. Here, they left the vehicle, and proceeded on foot, marching for hours in single file along narrow, winding game trails, the scorching sun beating down with an intensity that made the bush appear to almost 'crackle', and the ground felt red hot beneath their feet. The guide headed the column, with David following close behind, his rifle slung over his shoulder, and Bill bringing up the rear.

After several hours, it soon became apparent that Utopia always lay just over the next ridge! David began to show signs of impatience, but the guide reassured him that now they really were approaching their destination. Suddenly as they topped a rise the guide pointed proudly to a clump of green reeds amongst some large mud-stained boulders. The odd gnarled tree, also coated in red mud, its bark worn smooth by the rubbing of elephants, completed the picture. David and Bill paused momentarily and looked at the guide incredulously.

'This is Pundamilia spring,' he declared.

They continued to stare at him in disbelief, and then, Bill, who had an enviable ability to see the funny side of any situation, burst into uncontrollable laughter! David continued to look grim. The vision of his oasis faded, and he saw instead a stagnant seepage amongst a few boulders, the surrounding earth trampled bare and carpeted with a thick layer of fibrous dung. Looking down at the pool, he resorted to sarcasm, while the guide stood by with a slightly hurt look on his face, disappointed at the lack of enthusiasm his discovery had aroused.

'This is just marvellous!' David said, bitterly. 'We'll have swimming pools, waterfalls, and the lot – nothing could be more suitable for our headquarters! I think I'll build my shack under that magnificent tree!' he indicated a twisted looking specimen completely devoid of foliage, which set Bill off again, as he crouched to remove some gravel from his shoe. Suddenly, the laughter on his face was replaced by a look of horror directed at something behind David, and he literally hurled himself backwards, while the Waliangulu guide tore past at top speed, setting up what must have been a new Olympic record! David instinctively ran a few paces and then spun round to see an elephant towering over him: it had appeared like a genie from a clump of reeds, where it had obviously been lying. The animal swayed slightly, and then lurched a pace forward, and it was then evident that it could hardly walk at all, incapacitated by a foreleg swollen to three times its normal size. David raised his rifle and fired, and the elephant crumpled slowly at the knees and fell dead.

'Better still!' he remarked wryly, as he walked up to inspect the carcase. 'Just rounds off the day neatly! I'm sure Colonel Cowie will be suitably impressed with his new Wardens!' And at this, Bill could contain himself no longer, and collapsed with renewed mirth at the irony of the situation.

The cause of the swelling on the elephant's leg was very obvious. Sunk deep into the flesh, which was seething with bloated maggots, was a wire snare. David's bullet had brought weeks of suffering to a merciful end.

And so, his first day as Warden of the Tsavo East National Park drew to a close.

CHAPTER 2

The Waliangulu

I T very soon became clear that there was no hope of obtaining an adequate supply of surface water within easy reach of Voi and the railhead, to serve a headquarters. There was left no alternative but to explore the possibility of a well on the banks of the seasonal Voi River, and the site chosen was at a place opposite the illegal Waliangulu settlement, whose inmates watched the activity across the river with interest and a growing apprehension. Although the concept of a National Park was not to their liking at all, the Waliangulu were a carefree, optimistic people, with only two main interests in life – elephant and palm wine, the one being a means to the other, and it seemed unlikely to them that the two European 'greenhorns' and a few scruffy Rangers could alter their way of life very much.

The main tribe lived east of the Park boundary towards the coast, in scattered, isolated settlements, and the Ndololo contingent were an offshoot of these, who were rather frowned on

27

by the others due to their contacts with civilization and their liking for the bright lights of Voi. Since time immemorial, the Waliangulu had hunted game. They were thought to have been the original inhabitants of the Tana River, and because they were only a small tribe, and not sufficiently strong to stand alone, they had formed an association with some families of the powerful, nomadic Galla people. In exchange for the protection afforded them by the Galla, the Waliangulu provided meat and ivory.

Over the years, as the threat of inter-tribal conflict gradually diminished, the Waliangulu drifted from the custody of their protectors, and formed their own small settlements, where they lived, and still live, entirely by hunting.

Bows and poisoned arrows are used for this purpose. The poison, which is placed on the head and steel shaft of the arrows, and then wrapped in a protective hide covering, contains an extremely toxic glycoside known as oubain, which is derived from boiling the bark and leaves of a certain species of *Akokanthera* tree in water for about seven hours until a sticky, tarlike substance is produced. Sometimes, lizards, snakes and live shrews are thrown into the cauldron for good measure. The woody matter is then skimmed off the surface, and the mixture concentrated further by evaporation, until the required consistency is obtained. It is then packed in strips of maize husk, until ready for use, and this serves to shield the poison from the sun and rain which has a deleterious effect on its potency.

The poison is usually applied to the arrow shaft and head prior to an actual hunt. A great deal of secrecy and prestige is attached to the manufacture of the poison, and certain members of the Giriama tribe, who inhabit the coastal strip, are considered the foremost experts in this field. The potency of the poison deteriorates with age, but when fresh, it can kill an elephant within a couple of hours. It is active, however, only when introduced into the bloodstream, and causes death by upsetting the muscular contractions of the heart and arteries.

Although all *akokanthera* trees are poisonous to some extent, in most cases the berries are edible. For some extraordinary

28

reason, however, certain trees growing in the coastal belt are particularly lethal and the poison-makers are careful to select their material from these trees, which can be easily detected by the presence beneath them of birds and rodents that have died as the result of having eaten the berries.

The potency of the poison is tested in rather a brutal way. A thorn, which has been dipped into the poison, is jabbed into an unfortunate frog or lizard, and the time it takes to die is carefully noted. If the poison is fresh, a frog or lizard should succumb within a few seconds. It is said, also, that if an egg is pricked, and a little poison inserted, it will burst within half an hour if the poison is of good quality.

The meat from animals killed by poison is edible, and in no way tainted. Whenever possible, the arrow head is extracted from the carcase, for not only can it be used again, but the head normally carries the identification mark of the owner; a very important factor when several hunters may be operating in the same area, and an elephant might wander for several days before it eventually dies. When the locality of the carcase is revealed by the presence of vultures, the markings on the arrowhead establish positive proof of ownership of the animal. It is not considered 'cricket' amongst the poaching fraternity to remove the tusks from an elephant killed by another man's arrow, and retribution for this offence can be extremely harsh.

The poachers usually way-lay their victim by waiting by a tree overlooking a drinking place or well-used elephant path. When a suitable target presents itself, an arrow is fired into any part of its anatomy, and the spoor followed at leisure. Alternatively, the hunter might simply wait for the vultures to drop. The ivory is cut out of the skull, and usually sawed into manageable lengths to facilitate carrying, or it might be concealed nearby and collected at some future date. It is sold to 'middlemen' and fetches some two shillings per pound. It is then transported to the Coast and passed on at a handsome profit to unscrupulous Asian or Arab dealers, who smuggle it out of the country, often aboard dhows. Most illicit ivory and rhino horn ends up in the Far East, where it is turned into marriage bangles, ornaments and billiard balls in the case of ivory, or ground to a powder and used as an aphrodisiac, or a cure for rheumatism, in the case of rhino horn.

One has a certain sympathy for people who hunt for food for themselves and their families, and long ago this was what the Waliangulu, in fact, did. But the wholesale slaughter of animals, taking place when the Tsavo Park was created, was not for this purpose. The Asian and Arab dealers, while sheltering in the safety of their sordid shops in Mombasa's Old Town, actively encouraged the Africans to hunt elephant, rhino and leopard illegally, so that they could wax fat on the proceeds of the sale. The poacher, who is exposed to all the risks, is paid a pittance representing only a fraction of the true value of the trophies he brings in. It is these unsavoury dealers who are the true villains in the trade, for not only do they ruthlessly exploit the poacher, but they are responsible for the death of hundreds of animals every year. Seated behind some grimy desk, opulent and slippery, they control the destinies of dozens of harmless citizens, encouraging them to break the laws of the land, and corrupting officials in high places with their bribes. There is big money in the ivory racket – but only for the chosen few.

The Waliangulu were most active in that portion of the Park south of the Galana River, and along the Yatta escarpment. Further north lay the happy hunting ground of the Wakamba poachers, whose reserve borders the northern boundary. Like the Waliangulu, the Wakamba also make use of bows and poisoned arrows, but with the exception of a few individuals, they are not, on the whole, considered as fearless and efficient as their Waliangulu counterparts. Trophies from the north usually ended up in the same hands, albeit via a different route, and the proceeds derived from their sale in the same pockets.

During the first few weeks, while work was in progress clearing the dense undergrowth and levelling the ground in preparation for the base camp, David and Bill spent many hours chatting over a campfire to the Waliangulu, who often used to saunter over to satisfy their curiosity about what was going on. From these discussions, it became very clear that there was no love lost between the poachers and the dealers, for the Waliangulu were quite prepared to divulge general information, so long as it didn't become too personal. They were well aware of the fact that they were 'done down' by the receivers, and despised them heartily. And so, gradually, day by day, a picture of the poaching racket began to take shape.

CHAPTER 3

From Ndololo to Heartbreak Camp

ATTEMPTS to find water at Ndololo proved successful when a well was struck. Firstly, a kiln had to be constructed from sun-dried bricks, so that the bricks which were to be used to line the sides of the well, and prevent it from collapsing, could be properly fired. Making bricks was new to David, but the results were very rewarding.

Once the bush had been cleared, temporary workshops and stores were erected, and steps were taken to evict the illegal settlers on the opposite bank of the river. They were quite happy to go, because not only were they handsomely compensated by the Government for land they really had no right to be on in the first place, but were also allocated more fertile holdings elsewhere.

A disused locust track ran for a few miles through the Park in a south easterly direction, and Bill was given the task of opening this up, and pushing it on as a matter of urgency, by hand labour, towards a point on the Galana River near which stood a small conical hill marked on the map as 'Sala', the Waliangulu word for 'oryx'. I wonder how many people, driving along this same road today, now a broad all-weather highway, appreciate the sweat and toil that brought about its being. The work was fraught with every conceivable difficulty from the start. There was no road machinery in those days to ease the task; no bulldozers and graders, tippers or tractors; only the hands of a dozen men. The bush was extremely dense and had to be hacked away tediously every inch of the way, and to make matters worse, most of it was barbed with thorns. Great stumps and rocks had to be laboriously levered out of the baked ground. It was so hot between the hours of 11 a.m. and 3 p.m., that work had to be discontinued to allow the exhausted men to find what relief they could under any available shade. Bodies glistened with sweat from sun-up to sun-down, and on many occasions, elephant and rhino harassed the workers and sent them scrambling for the nearest tree. Hands got cut, toes got crushed, and many men succumbed to malaria. Water had to be strictly rationed, for it was more precious than gold. Every drop had to be carted from the Voi River forty miles away.

As the road slowly probed deeper into that sea of bush, the labour became more and more restive and reluctant to go so far from base. Many of them left, and others had to be recruited to bring the labour gang up to strength. The progress was agonizingly slow, and the extent of the poaching became more evident with every mile covered. There were signs of it at every waterhole where carcases of elephant and rhino littered the bush in alarming numbers. Live animals seemed to be extremely scarce, and David and Bill cursed the frustrations that prevented them from getting to grips with the problem, for the situation looked completely out of hand, and unless swift action was taken to curb the poaching, there would be no game left at all within the new Park.

Meanwhile, headquarters in Nairobi, impatient that the Park be opened to tourists as soon as possible in order to justify its existence and the money spent on it, urged that priority be given to this angle, which further frustrated the men in the field, who felt that all this would be to no avail unless urgent measures were taken to protect the game.

Finally, the one and only lorry, while carting water to the construction unit, was charged suddenly by a rhino and, in an attempt to dodge the animal, the driver collided instead with an enormous anthill, completely buckling the front of the vehicle and bringing it to an abrupt standstill. The rhino then got busy and added the finishing touches by driving its horn repeatedly into the mudguards and radiator, until the terrified Ranger who had accompanied the driver, shot the animal dead. It was this incident, the final misfortune in a long series of mishaps, that earned that lonely spot midway between the Voi and Galana Rivers the name of Heartbreak Camp.

The base camp back at Ndololo was soon completed and although it served its purpose well, it was never intended, in David's mind, to be the permanent Park headquarters. With a natural flair for farsighted planning, he had already set his sights on Mazinga Hill, and he could visualize the headquarters nestling on its eastern slope, with the Park laid out below, like a blanket, as far as the eye could see, and beyond. He knew exactly how he would build the headquarters complex, designed to cater for future expansion, commanding a pleasing view by being elevated above the surrounding flat bush country and able to benefit from any breeze in the hot, dry weather. There was, however, one serious obstacle to these dreams – Mazinga Hill lay outside the Park boundary, and was part of an adjoining Sisal Estate. David therefore made an approach to the owners of the Estate with a proposition for a land exchange, whereby Mazinga Hill could be ceded to the Park in exchange for a flat piece of country excised from a part of the Park south of the Voi River. After months of lengthy, drawn out negotiations, and legal agreements, the deal was clinched, and Mazinga Hill became a part of Tsavo East, and the site for a new permanent headquarters. Once again, the difficulty of water had to be overcome before work could be put in hand on this project, so another well had to be dug further upstream on the Voi River, and the water pumped and piped four miles, before work on the foundations could begin. Trees had to be planted to provide shade in years to come, building materials were difficult to acquire after the war and the stone had to be transported seventy miles from Taveta. Skilled local artisans were also few and far between, so that every aspect of the work needed close

and constant supervision, but nevertheless, the new headquarters quickly took shape. Houses for two Wardens were built
as well as offices, workshops, stores, labour and Ranger lines,
an armoury, storage tanks for water, and a well-equipped
garage to house the vehicles and machinery and where essential
repairs could be undertaken. In addition, three entrance gates
to the Park, complete with ticket office and staff accommodation,
had to be erected at strategic points along the main road, so
that visitors could enter the Park at any one of these places,
depending on the route they chose. And, all the while, the Sala
Road kept creeping towards its goal, the Galana River, and
still the carcases of numerous elephant and rhino were constantly being discovered by the construction unit. Tusk butts,
vacated hide-outs, ashes from fires, bones and broken calabashes
told their own grim story at every waterhole in the bush.
Patrols were constantly being sent out in pursuit of the offenders,
but the newly recruited Rangers were unfamiliar with their
surroundings and understandably reluctant to tackle large
gangs of poachers who were armed with poisoned arrows. The
poachers, realizing this fact, had little difficulty in evading
contact, and on several occasions, when cornered, had appeared
quite prepared to make use of their poisoned arrows, so that the
Rangers had been hesitant to close in.

As often as work permitted, David and Bill would accompany
the Ranger patrols, and carrying a few essential provisions and
a blanket, were swallowed up for weeks at a time in the immensity of Tsavo. They would follow elephant trails, drink the
muddy water from natural waterholes wherever they could
in order to conserve the supply they carried with them, which
could mean the difference between life and death in this thirsty
land. Tramping through the bush, they quartered the country
in search of an elusive enemy, who ran circles round them. And
when darkness enveloped them, and the sounds of an African
night followed the silent truce of dusk, they would halt where-
ever they found themselves at the time, snatch a frugal meal of
rice and bully-beef, and collapse into oblivion, rolled up in a
blanket under the stars, to sleep like the dead until another

dawn fired the eastern sky. Such excursions were seldom without excitement in the form of cantankerous rhino, or an old bull buffalo lurking in a thicket. Often, a possessive mother elephant necessitated hurried evasive tactics, sometimes a snake, attracted by the friendly warmth of the camp fire, slithered almost under foot, or simply, after a long, hot, thirsty day's march, they would find a waterhole dry, where they had hoped to replenish their *chargals*.

Living close to nature in this way, they came to feel the heartbeat of Tsavo, and slowly uncovered its secrets. They explored its depths, and had, at the end of such safaris, a better idea of what they were up against. And inevitably, being together day after day, facing common dangers and sharing past experiences, there was forged between these two men a mutual bond of friendship and respect, which was to stand the test of time.

It must have been on one such safari that David first knew of my existence. My brother, Peter Jenkins, had joined the National Parks service with Bill, and through him I had met Bill when I was still at High School. Bill's fun-loving, easy going, carefree nature, plus his fascinating occupation, had seemed the ultimate in any girl's dreams, and I had fallen in love with him.

I have always loved animals, having been raised on a farm in the highlands of Kenya, and belonging to a family who are all possessed of a deep feeling for nature. Much of my childhood was spent on safari in the Mara country, which was then a wildlife paradise. I cannot remember when first I heard a lion roar; it is something with which I have always been familiar, like the sighing of the wind, or the sound of a river; and so, the bush held for me no terrors, but instead an intense fascination. I remember the joy with which I greeted the news of my brother's new appointment, for this provided an opportunity to see more wild places, and more, too, of his attractive colleague. It was just accepted right from the time when I first met Bill, that one day we would be married, regardless of the words of caution delivered by my parents, who could see the differences in our two characters more clearly than we.

Similarly, it was through Bill that I first heard of David Sheldrick, and from the stories that Bill recounted, I visualized him as being a rather awesome character. Bill always spoke of him with deep admiration and respect as he related instances of his qualities of leadership, knowledge of nature, and his rather frightening efficiency. He also gave an insight into a dare-devil side to an otherwise serious and rather shy personality which, apparently, only became evident during occasional bouts of relaxation at the Voi Hotel after a long safari. I heard of wild parties at such times, where David entertained his fellow revellers with an odd assortment of parlour tricks, like chewing up glasses, blowing blazing paraffin from his mouth across the bar counter, provoking possessive husbands by brazenly flirting with their wives, and all sorts of other outrageous things, such as throwing darts at people's rear-ends, or plunging a fist clean through the bar window. He had even been known to try and stuff the Manager into the fridge! Bill then explained how they would have to return the following morning, full of remorse, to make their peace with the long-suffering hotel proprietor, and set about repairing the previous night's work, leaving everything as immaculate as before.

I had, in fact, met David briefly on one occasion, when visiting my brother and Bill, and I recalled a tall, aloof man, who on first acquaintance gave the impression of being faintly arrogant, and whom I had difficulty in picturing as the rogue in Bill's Voi Hotel stories. Little did I know then that in the years ahead he was to feature so prominently in my life.

As the Sala Road came within striking distance of the Galana River, it became obvious that elephant and rhino were being literally decimated by poachers, and that unless urgent counter measures were taken against the poaching gangs, very few of these animals would be left. Advance patrols came upon more and more carcasses strewn on both banks of the river, while vultures wheeled incessantly in the sky overhead. Also obvious was the fact that the patrols would not tackle the poachers unless accompanied by an Officer, so it was arranged that they be issued with ex-army ·303 rifles, in an attempt to instil more confidence.

It was also decided to push through to the river another disused hunters' track further upstream, which at present ended fifteen miles short of the Galana, and in connecting this to the Sala Road by means of a road along the river itself, not only would an interesting tourist circuit be opened up, but anti-poaching activities would also be greatly facilitated. This road was known as the Sobo Road, so named after an enormous solitary rock which stands beside the river at a point where the road emerged; 'Sobo' being the Waliangulu term for 'rock'. Again, the construction of this road was no easy task, for here the bush was, if anything, even thicker than that encountered on the Sala stretch. *Commiphora* trees growing so close together that their branches were interwoven, formed an almost in-penetrable barrier, beneath which the *Sansevieria*, or wild sisal, struggled for standing room in competition with a tangle of other vegetation. Half the labour gang was again laid low with malaria; there were strikes and endless complaints to deal with plus serious vehicle breakdowns; and to cap everything, an unfortunate elephant with an arrow shaft projecting from its stomach, collapsed and died within a few hundred yards of the camp, which made life even more unpleasant. Then came the discovery of a cooking pot close by, a torn, bloodstained loin-cloth, and a bow and quiver full of arrows lying on the ground. Staring at these remains, aghast, one of the Waliangulu labourers recognized them as having belonged to his father, who had failed to return from a recent hunting trip, and who now in all likelihood never would, having probably fallen victim to a hungry Tsavo lion. This discovery provoked yet another serious interruption to the work. The labour, crowding around, their eyes wide with horror, chattered excitedly, and felt even less inclined to remain in such a remote place. But in spite of everything, the road progressed, and when it was approaching the river, David planned an all out offensive against the poachers.

An armed patrol, led by a veteran Ranger, was detailed to go to a point on the river near the eastern Park boundary, there to wait in ambush, while David and some other Rangers, starting at Sobo, worked slowly downstream, hopefully driving any poachers operating in the area towards the ambush.

The sweep hadn't been under way very long before a faint chopping sound could be detected in the distance, so the patrol deviated slightly in order to investigate. As they drew closer, odd bursts of song and loud whistling accompanied the chopping, and very soon the originator of the sound came into view as he merrily explored a likely looking crevice in the trunk of a tall acacia tree in the hopes of finding honey. Leaving the Rangers concealed in the bush, David strolled quietly up to the base of the tree, and stood looking up for several minutes, while the man, oblivious to everything but the prospect of honey, continued with his work. David spoke sharply.

'Get down!' he said.

The chopper was poised in mid air, the whistling ceased instantly, a look of utter disbelief and shock, which gave way to one of horror, came over the poacher's face, as the chopper fell to the ground, followed by the man himself, who landed with a dull thud. Miraculously, he failed to break his neck and was hastily pounced upon and pinioned to the ground by the jubilant Rangers, whose morale had rapidly soared with the chance of success. It was a few minutes before the man was in a fit state to speak coherently, but he finally stammered that he was one of a gang of five hunting in the Park, and readily agreed to guide the patrol to the hide-out, which consisted of a hollowed out bush on the river bank. A careful scrutiny through binoculars as they approached this place, revealed that the other members of the gang had yet to return, for there was no sign of any activity in the hide-out. The patrol therefore hid within and around the hide-out to await their quarry.

Some hours later, voices were heard on the far bank of the river, and soon four men came into view, all carrying bows and poisoned arrows. Two lingered to drink from cupped hands at the water's edge, while the other two strode up to the hide-out. The first man reached the entrance to the hide-out, and then his companion was startled to see him disappear rapidly, as a white arm shot out and dragged him inside. He himself was roughly seized and dragged inside as well, and the two remaining poachers wasted no time, but began to sprint in a shower of spray across the river. David emerged from the hide-out, quickly levelled his rifle, and placed a bullet in the water beside one of the fleeing men, who promptly fell flat on his face with terror, giving the Rangers an opportunity to catch up and overpower him. The remaining poacher, who summoned an added burst of speed he never knew he had, was conscious only of the sound of the shot and his comrade having fallen. He must have run most of the way to the Park boundary, obviously thinking the worst, and just as he was about to cross over, no doubt heaving a sigh of relief at having escaped with his life, he was seized by the Sala ambush. They were surprised to find him in a state of near collapse, and even more surprised when they finally managed to extract from him a halting account of what had taken place. He told them that two members of the gang had been arrested and one had been shot! But no one was more surprised than this poacher, when, two days later, the deceased turned up with the other prisoners and the patrol, very much alive!

Prisoners were always a problem for they had to be constantly guarded and fed, which was not always easy towards the end of a long and gruelling foot safari, when the Rangers were tired and short of food themselves. When on the march, any prisoners were usually handcuffed together in pairs, so that they could be recaptured without difficulty if they attempted to bolt, but this also posed problems when dangerous animals were suddenly encountered. On one of the early anti poaching patrols, a number of poachers had been arrested, and they had agreed to lead the Rangers to the hide-out of another gang operating in the same area. While passing through some thick bush, the party was suddenly charged by an elephant. It was every man for himself to take what evasive action he could, including the prisoners, two of whom were brought to an abrupt halt by their handcuffs when they passed either side of a tree. Luckily, David

The first headquarters of the Tsavo National Park, established in 1948

Two Rangers climb an anthill to try to get their bearings in the dense bush

'Miles and miles of nothing'. Bill looks out over Tsavo East

Roads had to be hacked out of the bush by hand

noticed their predicament, and thinking that they would be killed by the enraged elephant, who was rapidly gaining on them, prepared to shoot the animal should the necessity arise. Very fortunately, the elephant crashed off into the bush, swerving away at the last moment, but leaving two very shaken prisoners trying desperately to disentangle themselves! Following this incident, it was difficult to resist the prisoners' urgent pleas that their handcuffs be removed while walking in the bush, but at the same time the risk of their escaping had to be borne in mind.

Interrogating prisoners was a very time-consuming business, and required a subtle approach to extract the maximum amount of information. David and Bill would spend many hours chatting to them unconcernedly over a camp fire about things of mutual interest, which bore no apparent relationship to poaching at all. They would discuss elephants, recount unusual experiences, enquire about a known relative, or simply chat about the weather. In this way, poachers usually became more talkative, and a wealth of information was slowly extracted, which was carefully documented and filed for future reference in what was called 'The Rogue's Gallery', a small cabinet that contained a card on every known poacher, where information relating to him, however trivial, was recorded against the day when he was finally brought to book. Each prisoner also had his portrait taken on capture, and this completed the card, so that he could be easily recognized should he be caught poaching again in the Park at some future date.

Often an anti-poaching patrol would be many weeks in the bush before returning to base, and captors and captives came to know each other quite well during this time, establishing by the end of a long safari a friendship based on mutual respect and common interests. Sometimes, a prisoner who proved very cooperative and volunteered useful information, was released with a warning, or else enlisted as a paid informer. And very often, too, having been convicted and completed a prison sentence, a poacher would appear at the headquarters, greet David like a long lost brother, and enquire about the prospects of a job.

David always did his best to find employment for such people, usually with professional hunting firms who needed trackers or gunbearers, and in this way ensured that a potential elephant killer was kept otherwise occupied, for a period at least! There was never any feeling of animosity attached to the capture of a poacher. They regarded it as in the luck of the draw, and were quite prepared to cheerfully accept the consequences without resentment in a manner befitting a gentleman.

By far the most tedious aspect was the prosecution angle, for sometimes what first appeared to be a clear cut case, took an unexpected turn, and could drag on for days, taking up a lot of valuable time that could be better spent elsewhere. The intricacies of the law very often confused the simple poacher, and there was an occasion when the proceedings went something like this:

'You are charged with killing an elephant in the National Park at Lali on the 28th of January this year,' said the Magistrate.

'Yes, I killed the elephant,' replied the accused.

'Nevertheless, you are not guilty of this offence until the prosecution has proven the fact,' said the Magistrate.

At this, the poacher looked blank. The Magistrate turned to an interpreter, and instructed him to explain to the accused that, in the eyes of the law, he was not guilty, and the prosecution would be required to call witnesses.

'But, I did shoot the elephant!' exclaimed the poacher, growing quite heated. 'Do you think I could miss it when I was standing only ten paces away?'

'No, no!' interrupted the Magistrate, 'as far as this court is concerned, you are not guilty until it is actually proved otherwise.'

The bewildered poacher looked entreatingly at David, who was standing behind, as though to say 'Is this man mad?' It was difficult to explain to him that under certain circumstances, he would have been justified in shooting the elephant; in defence of his person or his crops, because, in framing the charge, the police had inadvertently omitted the word 'unlawfully', and as a result of this error, the onus was on the prosecution to prove that the elephant had, in fact, been 'unlawfully' killed!

The accepted method of counteracting poaching in most established National Parks elsewhere had always been to operate from a series of outposts situated at strategic points, but in Tsavo this system was proving a dismal failure. Several Ranger Posts had been built in the Park, at Sala, Tsavo, and a place called Kimethena on the northern boundary where it bordered the Wakamba country. Two or three Rangers were permanently stationed at each, whose duties were to patrol a section of the Park each day, but more often than not, they were afraid to venture out, and spent the day whiling away the time at the Post, gaily reporting that they had found no evidence of poaching in their area. When the carcases of two poached elephant were found not more than 300 yards from the Sala Post, the Rangers being quite unaware of their presence, and when a gang of poachers actually staged a raid on the Ranger Post and made off with various items of equipment, it was quite obvious that a better system would have to be devised if the poaching was to be contained. The Rangers at the Kimethena Post had even concluded an agreement with some Wakamba poachers whereby, in exchange for permission to collect wild honey in the Park, they would be given free beer! However, one other incident was responsible for bringing matters quickly to a head.

A dead elephant had been seen beside the Sobo Road, so a party of Rangers were sent out to assess the cause of death, and if need be, to lay an ambush by the carcase. When they arrived on the scene, they found that the tusks had been taken, so they followed the spoor of the poachers, which led them to a second dead elephant and four men actually in the process of extracting the ivory. As soon as the poachers became aware of the presence of the patrol, they fled, but one man, who at some time had been lamed by a crocodile bite in the foot, found that he was unable to keep up with his comrades. The leading Ranger closed in on him and managed to throw him to the ground, where they grappled together. During this desperate struggle, the poacher drew a knife, and stabbed the Ranger twice in the heart, before leaping clear and making off. By this time, the other Rangers had caught up, and ended up by shooting the poacher dead. There followed an anxious discussion about what to do, and it was decided that one man should set off immediately to report the matter, involving a walk of thirty miles; while the others would carry their fallen companion to a point on the Sala Road, leaving the body of the poacher behind.

David received the news just before nightfall that same day, and having hurriedly reported the incident to the police, went with them to the scene of the tragedy. They met the patrol on the road, guarding the body, and while the police conducted their investigations, David and some other Rangers set out immediately in an attempt to intercept the members of the gang that were still at large. Assuming that the poachers had come from the main Waliangulu settlement of Kisiki-cha-Muzungu, ambushes were set near the Park boundary on every elephant path leading in this direction. Most unfortunately, however, it was discovered later that the gang had, in fact, made for the Garbiti settlement, which lay in another direction, so contact was not made and the long night's vigil was to no avail.

The next day, the body of the poacher was recovered and taken back to Voi, where David promptly sat down and wrote an urgent letter to the Director of National Parks, seeking authority and the necessary funds to recruit and train a special anti-poaching unit, resolving as he wrote this letter, to avoid recruiting local people for this particular aspect of the Park's work, but to concentrate instead on training men from the nomadic tribes of northern Kenya, who were not only fearless and best suited to this task, but were also used to living under arid, harsh conditions, and were skilled in bushcraft and the art of tracking. In future, all anti-poaching work would be undertaken by these men, who would be known as the Field Force, and the ordinary Rangers who had been recruited locally, would only man the entrance gates and deal with tourists.

The necessary authority was, in time, given, so David sat down to devise an intensive training programme, based on military lines, while Bill set off to recruit Turkana, Samburu,

Somali and Orma tribesmen from their own homelands in the Northern Frontier.

The first lorry load of recruits duly arrived, and it was difficult to imagine that they would ever be a disciplined force, for they were plastered in red ochre, clad only in a thin loincloth, carried spears and couldn't even speak Swahili. Their arrival in Voi caused quite a stir amongst the local populace, whose fear of such people lingered still, a relic of the dark days when the Masai used to raid other tribes, seizing their stock, abducting their women, and leaving death and havoc in their wake. But, after three months' intensive training, these same recruits were transformed and ready to go into the field, the embryo of what was to become an extremely effective striking force which would act as the Blue Print for every National Park in East Africa.

CHAPTER 4

The Initiation

I T was a great day when Tsavo East acquired an old Cater-
pillar Grader, even though this veteran machine had seen better
days, and had been reconditioned following war service in the
Philippines. Nevertheless, the construction of roads was now
made a lot easier.

The Grader became Bill's particular toy, and he greatly
enjoyed driving it himself whenever opportunity permitted. It
was even given a name: 'Daphne' was emblazoned on its side,
and when Bill told me proudly that he had named the Grader
after me, I tried very hard to look suitably honoured!

Bill on safari was quite an education. He had acquired an
assortment of livestock during his time in Tsavo East: two goats
from a Waliangulu friend, some strange looking chickens, whose
necks were quite devoid of plumage, giving a nude appearance,
and a mongrel pup. He was also an enthusiastic amateur on the
guitar, and had as well a passion for cowboy records which
he played incessantly on a squeaky old gramophone that
required constant winding. And so, whenever he set forth on a
safari, amongst the camp equipment and tentage piled high on
the back of the lorry, were the two goats, all the chickens, the
dog, the gramophone and the precious guitar, and these items

46

always went everywhere with him. At each halt *en route*, where copious quantities of tea were swallowed, the entire menagerie, including all the chickens, would take the opportunity to stretch their legs and explore their new surroundings, but as soon as the lorry started up, there followed a wild scramble to get aboard again. All the livestock were quite resigned to their roving existence, and seemed to take each adventure in their stride.

Bill's camps were usually to be found in the most unlikely spots. Whereas most people would have selected a nice shady open clearing, Bill was never happier than when buried in the densest thicket. The tent usually housed the records, the gramophone, guitar, goats and some scant provisions, while he, himself, was quite content to sleep in an old hammock slung between two trees. The chickens and dog retired to the lorry at night. It was usually quite easy to home in on the camp, if one had occasion to call on Bill, for invariably a scratchy cowboy ditty, accompanied by the strumming of the guitar, and the crowing of the rooster could be heard some distance off.

I shall never forget my initiation into Bill's camping habits! I had accompanied my parents and family on one of their visits to my brother, Peter, who was based in Tsavo West, where we had spent a very pleasant Christmas with him in his camp at Mzima Springs. My father had trained Peter well, and his camp was immaculate with every comfort provided. The actual site itself was idyllic and had obviously been chosen with great care; the tents nestled under shady yellow-barked acacias, and had a pleasing outlook; the camp beds had comfortable mattresses; a drum with a fitted shower attachment, hoisted into a tree and concealed by a brush fence, provided excellent ablution facilities, and each tent had its own camp basin, table, and chair, with soap provided and a clean towel pegged onto the guy rope with a clothes peg. The cook produced the most sumptuous fare cooked over an open fire, and an old paraffin tin with a fitted shelf and lid served very adequately as a safari oven. Peter's camp was a replica of those I remembered from my childhood safaris in the magnificent Mara country.

We had also received a very pressing invitation from Bill to visit him in Tsavo East, and after a lot of persuasion, I succeeded in pressurizing my father into accepting this offer. The short-comings of Bill's camp, in my father's eyes, would perhaps not have been so apparent had we not just come from the beautiful surroundings of Mzima, and certainly, the misfortunes that beset my father there were not of Bill's making. We had some difficulty in locating the camp, to begin with, not being con-versant with Bill's camping idiosyncrasies, but eventually in our wanderings we happened to meet him, and were led to it. By this time, my father was feeling the heat, and not in the best frame of mind. His face clouded further when we reached the camp, and found it surrounded by very thick undergrowth which gave it a claustrophobic atmosphere. Two tents, Bill's hammock, a table and some chairs and wooden boxes completed the scene.

A series of unfortunate events then took place. Firstly, an army of safari ants had got into my parents' tent and were responsible for a great amount of discomfort before they were repelled, then dull thumping sounds from the back of the tent heralded the demise of a puff-adder found there by the cook. The evening meal was invaded by stink bugs, whose permeating stench polluted the soup, making it inedible, and when my father came to retire for the night, he found a large and extremely menacing looking scorpion on his bed. That night the lions roared all round the camp, and my mother kept getting up to see whether we were all present and correct. It was far too hot to sleep with the tent flaps closed, and so we barricaded the entrance to the tents with chairs and boxes, which we hoped would deter any would be man-eater. And throughout all this our host slept soundly in his hammock, quite oblivious to the disturbance caused by both his guests and the lions. When we bid farewell to Bill after a couple of days, I got the impression that camping with Bill would be an experience my parents would not wish to repeat too frequently!

The need to provide more watering points for the game was

very urgent, for their numbers would never show a noticeable increase while the limiting factor lay in available food during the long dry spell, when animals were forced to concentrate on the few sources of permanent water. David was very anxious to make a start in the right direction by attempting a dam on the Voi River, so that the water that raced down it during periods of heavy rain, could be impounded and utilized throughout the dry season. But he didn't receive much encouragement from the District Commissioner, who adopted a bored, rather patronizing attitude towards the concept of National Parks, regarding it as an expensive, harebrained idea doomed to eventual failure. Added to this, was the fact that the District Commissioner, who was a friend of David's, quite enjoyed getting a rise from him on occasions when they dined together. Convinced that there was a place for National Parks in Kenya, and that wildlife would one day prove a valuable economic asset to the country, David was always quick to defend the cause.

'I haven't seen you for a week or more,' the District Commissioner said one evening. 'Where have you been?'

'I've been busy taking levels near Bwoka. We are anxious to dam the river before the onset of the rains,' David replied.

'What on earth for?' asked the District Commissioner, feigning surprise.

'Well, there is plenty of grazing available in the area, which at the moment can't be utilized in the dry season, owing to the lack of water.'

'Don't say you are proposing to open the Park for cattle!' exclaimed the District Commissioner with obvious amusement.

'No, of course not,' answered David, trying to conceal the irritation he felt. 'We want the water for game.'

'Oh, I see,' smiled the DC. 'But what makes you think the dam will hold water? Remind me to give you the Public Works Department Reports where you will see that the soil in this area is porous and unsuited to dam making.' He paused. 'A pity you didn't ask before about this – it might have spared you some trouble.'

'The dam at Mwatate seems to hold water,' remarked David, bristling.

'Yes, indeed,' mused the DC. 'I suppose if it is kept topped up by the river you might keep a puddle, but I feel the money could be channelled into something more constructive, rather than encourage game to spread. We already have enough trouble protecting crops.'

By now, David would be beginning to get heated. 'The success of this Park is dependent upon building up the stocks of game. You can't expect tourists to visit these Parks unless there is something to see!'

'My dear boy!' drawled the DC. 'A pipe-dream! Tourists won't come to a God-forsaken place like Tsavo to gaze at a lot of destructive and dangerous beasts. Anyway, quite apart from all that, it's not feasible to sit on a vast block of land for the exclusive use of wild animals, with land hunger so acute in Kenya. Sooner or later it must be opened up for settlement.'

'Not if we can prove that a National Park is the best form of land use in marginal areas,' snapped David.

The DC chuckled maddeningly. 'It might interest you to know that the population here will have almost doubled within the next twenty years, and you don't suppose the African will sit back and watch you breeding wild animals to destroy his crops and kill his goats! No, the Government will have to provide land for all these people, and your Tsavo Park will go by the board!'

By now, David would be finding it difficult to remain polite to his host. 'What! Destroy a priceless heritage in the process?'

'Hungry people are not going to worry about that! To me it's all so obvious, I cannot understand your blindness. You are surely not a complete moron, and you have your life before you. Why not start looking for another job? There is no future in Parks, I assure you.'

Arguments such as this not only infuriated David, but they also depressed him, and sometimes, when things seemed to crowd in on him, he wondered if he was, in fact, wasting his time, and whether he could ever make any impression on Tsavo. But

50

he was heartened considerably when despite the DC's gloomy prediction, the first small dam at Bwoka did hold water.

Much encouraged, a more ambitious project further up-stream at a place known as Kandecha was attempted, and this proved successful as well, impounding a sufficient quantity of water to last through a good deal of the dry season. But it was still not permanent, and so the possibility of yet a third dam further downstream was investigated. After taking very careful levels, it was found that by throwing up a wall of earth thirty feet high at a point below Bwoka, it would be possible to create a lake in this arid land.

The actual construction of this project was well beyond the capability of Tsavo East's most recent acquisition – a D4 bulldozer – and so the services of a contractor had to be enlisted. For weeks a cloud of dust hung over the site, as the tractors levered up enormous quantities of soil and piled it across the mouth of the catchment area. Great scars seared the earth where the soil was excavated, but finally the wall was completed, and when the storm clouds gathered and poured out their precious contents over the Teita Hills, bringing the Voi River down in a red angry torrent, everyone watched a lake appear with immense pride and satisfaction. It was called 'Aruba', which means 'elephant' in the Waliangulu tongue.

There was great excitement when the first few visitors pur-chased their entrance tickets at the gate. David, anxious that they be satisfied customers, went along too to ensure that they got their money's worth. But in those days, the bush on either side of the narrow roads was so dense that unless an animal crossed the road in front, or actually stood in the centre of it, one could pass within a few feet of it and be quite unaware of its presence. Also, unless one knew where to go, it was possible to travel 100 miles and more, and not even see a dikdik, so David's fears were not without foundation. Platforms overlook-ing favourite drinking places were constructed in trees, so that visitors could have a vantage point from which to photograph game. A small tree house was also erected at a place called Maji Chumvi, where elephant and rhino congregated to lick the mineralized soil from a seepage below the tree, and this, too, proved popular with those early visitors to the Park. One drawback was that people could find themselves marooned up the tree a lot longer than they had bargained for, which could prove inconvenient.

Very soon, there was talk in Nairobi about the necessity to provide Lodges in the Parks, where visitors could be accommodated in comparative comfort. One such establishment already constructed in Tsavo West had proved a great success, but the scenic beauty and pleasant surroundings to be found there, plus the many sources of permanent water, made the choice of a site a lot easier. In Tsavo East, suitable places were practically non-existent, and one's choice was further restricted to permanent water. The Galana River was subject to extremely severe flooding during the rains, as was the Voi River to a lesser extent, so it was eventually decided to site the Tsavo East Lodge on the shores of the new lake. There were, however, no shade trees, so these would have to be planted, and it would therefore be many years before the surroundings at Aruba Lodge could begin to compare with those of Kitani in Tsavo West, if ever. But a start had to be made somewhere, so the bush was cleared and six self-contained thatched cottages were built. Spindly little seedlings that would one day grow into tall evergreen *Melia* trees were planted to provide much needed shade in the future and lend a touch of beauty to an otherwise stark, windswept scene. Melia trees were chosen for a very good reason; they happen to be one of the few species that are not touched by elephant.

Throughout this period, and in addition to their normal duties, officers of the National Park were very often called upon to deal with elephant, and occasionally other animals, molesting neighbouring settlements. Two establishments seemed to suffer most trouble – the Military Cantonment at Mackinnon Road, some forty miles away from Voi, and the Sisal Estate adjoining the Park. In the interests of good relations with the Park's neighbours, an attempt was usually made to go some way towards appeasing demands for action by shooting the most

Elephants at a water-hole

Rhino with a small calf beside her

persistent troublemakers, which more often than not turned out
to be bull elephants, usually more daring than the cow herds.
In the case of the Military Cantonment, the elephants were
attracted by the presence of water, which was used to irrigate
the lawns and gardens within the compound.

One persistent group of bulls, despite having been chased out
the previous day, was back again the next morning, and it was
not long before large inquisitive crowds began to collect. The
elephant, finding themselves cut off, began to panic, and finally
stampeded. In the ensuing mêlée, with people fleeing in all
directions, and elephants screaming and trumpeting, one animal
fell into a deep ditch, breaking a leg, and was injured yet further
by being trampled on by those following. Yet others still some
way behind were turned by the crowd, who by this time were
beyond the control of both the civil and military police. Panic-
stricken, the elephants charged back towards the camp, and
huddled in a terrified group, completely surrounded by an
excited and chattering wall of humanity, until one soldier, in
an attempt to capture the scene on film, approached too close,
and was seized and flung high into the air. He must have died
instantly, for every bone in his chest was crushed like so many
matchsticks.

At this point, the Police Inspector in charge of Mackinnon
Road police station got to hear of the incident, and deciding
that he must do his stuff by dispersing the crowd, hurriedly
crammed six police constables into the mesh enclosed back
section of a pickup, and raced to the scene where he pulled up
in a cloud of dust. By this time, a section of the crowd was
again in the process of converging on the elephants, so the
Inspector, whose devotion to duty on this occasion overcame his
well-known terror of elephants, leapt from his car and ran
towards the crowd, shouting to them to move back. His warning
proved unnecessary, however, for at that very moment the
elephants decided to charge the crowd, who broke up and stampeded wildly in all directions, leaving the horrified policeman
confronting an avalanche of elephants.

The Inspector, who was a man of impressive proportions,

and amongst other things the Police Heavyweight Boxing Champion, decided that discretion was the better part of valour at this stage, and also took to his heels, but feared that he would be overtaken by the enraged elephants long before he could reach the safety of his vehicle, so dived headfirst into an eighteen inch culvert underneath the road instead. His terror of elephants must have clouded his judgement somewhat, for he found his head and shoulders firmly wedged in the entrance to the culvert like a cork in a bottle, but could make no further progress despite the frantic pumping of his legs, which were working overtime in an attempt to push himself further in! Meanwhile, six panic-stricken constables found themselves in the unhappy predicament of being unable to escape, being still locked in the cage at the back of the vehicle, and were yelling loudly for help. They must have been extremely relieved when the elephants turned suddenly, and retreated into a clump of bush.

The Inspector, on the other hand, was unaware of this development and continued to kick feebly in the mouth of the culvert.

It wasn't long before the crowd saw the humorous side of the situation, and a great roar of laughter accompanied by cheers and yells of encouragement went up. After some minutes, the Inspector reversed sheepishly out of the culvert, and proceeded to dust his starched jacket, obviously suffering from a loss of dignity, and feeling acutely embarrassed.

He was at the Mess, fortifying himself with double brandies when David arrived, having been summoned from Voi to deal with the elephants. The Inspector required a good deal of persuasion before he would agree to show David where the elephants were to be found.

'Everyone's gone mad round here!' he said heatedly. 'They're all bent on suicide, and I don't want any further part in it, thank you!' But, after a couple more stiff brandies, he finally reluctantly consented to accompany David, possibly reassured by the sight of a heavy rifle.

They found the terrified elephants bunched together in a small patch of bush, and summing up the situation at a glance,

David realized that it would be dangerous to attempt to drive the animals in any direction, for even with the aid of dogs, the police had been unable to clear the perimeter of onlookers. Any further harassment of the unfortunate animals would only lead to another stampede and possibly more casualties. He decided that there was only one course of action under the circumstances, and that was to stand by until dusk, when the perimeter road could be placed out of bounds, and the elephants could under cover of darkness find their own way out of the compound and return to the shelter of the Park. Much to the relief of all, this prediction proved correct, and by the following morning all the elephants had managed to escape.

But this experience did not teach the elephants a lesson, and they continued to cause trouble at Mackinnon Road periodically until several ended up by being shot. There was another occasion that provided good entertainment for the troops when a lone elephant chased the Brigadier around the compound, but not all such happenings had their share of humour, though, and the occasion when three elephants fell into a narrow but deep washaway was tragic in the extreme.

A message from the Headmaster of an African school on the slopes of the Teita Hills, read like this: 'We are invaded by a big crowd of elephants – we cannot learn. Please inform the National Parks to do something to help today before they do more harm.'

David hurried to the school, where a most distressing sight greeted him. A terrified elephant, being tormented by hordes of people who lined the sides of the washaway, was standing on the bodies of its two companions. All three had apparently fallen into the washaway during the night, and in their frantic attempts to clamber out, two had been trampled to death. It might have been possible to break down the banks of the washaway and allow the animal passage, but the area was very thickly populated, and the animal so panic-stricken, that there was a very real danger of someone getting killed as the result. Reluctantly, therefore, the only alternative was taken, and the animal's suffering was swiftly ended.

CHAPTER 5

Events leading up to
the Anti-Poaching Campaign

As the Park was more thoroughly explored, it was discovered that the poaching in the extreme North was also very serious. The only means of access to this remote area lay in a long, roundabout route through the Wakamba Reserve, for the Galana River successfully isolated it, cutting Tsavo East into two distinct halves. Quite obviously, the next most important requirement was to construct a causeway across the river to link the two halves and establish improved communications.

This presented a great challenge, for the Galana is not only some 300 yards in width, but is also subject to heavy flooding twice a year during periods of heavy rain, and diverting the flow to enable the concrete to be laid presented great difficulty. A suitable rock seam was located at a point just above Lugard's Falls, which would form a base upon which the foundations could be laid.

The term 'falls', when related to Lugard's, is probably not a very apt description. The river, elsewhere wide and sluggish, is suddenly constricted at this point and funnelled down to a very much lower level through a deep, narrow chasm, no more than eight feet across. The enormous volume of water, which boils angrily, hurling itself through this fissure, has, over the years, worn the rock to contemporary artistic shapes that give the falls a sinister, unearthly atmosphere. Incredible rock formations are made even more beautiful by the embedded deep red fragments of garnet, and the delicate hues of compressed layers of a darker tone within the basement complex. Below the falls, the river opens out once more, and the water, as though spent by the force of its descent, lingers there to gently lap the sandy edges before slowly proceeding on its way, while indolent crocodiles, engorged on the fish who battle to climb that boiling torrent, laze peacefully and quietly on the sandy spits, and stately herons silently wade the waters.

An eighteen foot fault in the rock seam near the south bank necessitated the construction of a bridge to span the gap, and great quantities of sand had to be removed to expose the rock base for the buttresses. It was necessary for men to toil day and night in shifts to bale out the water which seeped in incessantly through the sand, to enable the concrete foundations to be finally laid.

In addition to diverting the flow of the river, deep cuttings had also to be excavated through the various quite sizeable islands in midstream. Ballast for the project was broken by hand labour, for there were no luxuries like cement mixers and stone crushers in those days. Such items were way beyond the Park's meagre budget. This work was closely and constantly supervised by a very able Assistant Warden who had been newly appointed, John Lawrence. It proved to be a long, drawn out project, though, for every time the river came down in flood, work had to be abandoned until the water subsided again. The causeway took a year to complete, but it represented a great stride forward in the development of Tsavo East, bringing the vast neglected Northern Area into the fold, and although the causeway has since been widened, the original structure stands to this very day; a credit to the man who was responsible for its construction.

The poaching in the Northern Area, carried out by Wakamba tribesmen, was most serious along the Tiva River. This was already known from one or two exploratory safaris undertaken prior to the completion of the causeway, when an approach had been made from the north along an old bushtrack cut by MacArthur many years before. On one of these occasions, Bill had become hopelessly bogged in the sand while attempting to cross the dry Tiva riverbed, and had battled desperately for several anxious days to extricate the lorry, his supply of water so short that he had to be sure to guard the precious reserve in the radiator from the thirst-crazed men by sleeping in front of the lorry at night. Had he failed to get the vehicle free, he would have had to face a long trek through waterless country with only what was left in the lorry radiator, and he therefore had to conserve it at all costs regardless of the desperate desire of the moment.

Unknown to him, the plight of his party was observed by three Giriama poison sellers, two men and a woman, *en route* to the Wakamba country to dispose of their wares. They had made for this point on the Tiva River where they had hoped to replenish their water containers from a small spring they knew of in the area. Unknown to them, this spring was, however, completely dry. They had already crossed sixty miles of waterless country and had a further fifty miles to cover before their destination was reached.

Seeing the National Parks vehicle stuck in the sand, they decided to conceal themselves in the bush until it had passed on, having no wish to be caught at this stage of the game. Bill, meanwhile, oblivious of their presence, continued the struggle to get the lorry out, jacking it up, laying down brush and logs beneath the wheels, digging away the steep banks and permitting his men only the odd sip of water; and all the while, from the fastness of the bush nearby, the three poison sellers were as anxious for success as he so that they could get access to the water they thought they would find there. Finally, on the evening of the second day, the lorry managed to grind its way clear of the drift, and with a silent prayer of thanks through parched swollen lips, Bill and his party faced the long slog back to the Galana River, praying fervently that no further mishap would delay them. They hadn't gone far along the track when he happened to notice the footprints of the three Giriama and after pausing briefly to examine them, decided that his own circumstances were too desperate to allow for investigation. So, he pressed on with all haste and proceeded to the Galana. Many weeks later, he heard the sequel to this story.

Finding the spring at the Tiva crossing completely dry, the luckless Giriama had no alternative but to attempt to reach a place called Mutha, fifty miles distant. It was not long, in that searing heat, before one of the men collapsed by the side of the track and declared that he was unable to go any further. The woman volunteered to stay beside him, while the remaining man went in search of water. It must have been with great misgiving that she watched him disappear into the bush, and as darkness fell, and the dying man continued to plead deliriously for water, there was still no sign of the other man's return. She must have been demented with fear during the course of that long night, and one can imagine her crouching beside her companion until he died, staring into the darkness, possibly having to listen to the eerie call of the odd hyaena, or the roar of a nearby lion. At first light she abandoned the body and stumbled into a village in the Wakamba Reserve two days later, unable to speak and almost dead. She had made it – but, only just!

A party set out later to try and locate the man who had gone in search of water, but they found only his body, lying not very far from where the first man had died.

Soon after this the Mau Mau rebellion began, and a State of Emergency was declared throughout the country. Bill fell into the category for call up, and left to undergo a period of military training followed by service with the Kenya Regiment. David, too, would have liked to get back into uniform, but the authorities would not release him, so he had to content himself with harassing poachers instead, although he did take advantage of leave periods to soldier briefly with the mounted section of the Security Forces.

By this time, his marriage had broken up, and Diana had left with their two small children.

He missed his children acutely, and there followed a period of deep depression. There were times when he almost hated Tsavo: the poaching problem allowed him no respite, the rains were so unpredictable that they became an obsession, and one found oneself continually glancing skywards trying to detect any trace of hope, and becoming more and more depressed as one hot breathless day passed into the next. Funds were limited and always short; there was so much that needed to be done, but no money to do it with. He felt as though he were floundering in a morass of quicksands where no progress was possible.

And then one day as he sat on Mudanda Rock, and gazed upon some 300 elephants frolicking in the pool below, he knew that his efforts were not in vain, and that his reward would be to one day see this Park filled with wild animals. He must strive to create a Park that would equal others in Africa, and so earn on its own merits an enduring place in the country. The depleted game stocks must be nurtured back to healthy numbers, encouraged to multiply by the creation of suitable conditions. It would be a long, slow process, but what thing more worthwhile could one do with one's life than preserve for posterity that closest to one's heart? David found solace alone in wild places, and slowly the breach left by the disintegration of his marriage healed and closed.

Elephants had always had a particular fascination for David, as they have for many people who have the opportunity of getting to know them well. So huge, yet so vulnerable, so powerful yet so gentle, so wise, so tolerant and so vitally important to the ecology of Tsavo. They were the key to the proper management of the Park, the main link on which all other species were hinged. They were the heartbeat of Tsavo, and in order to understand many aspects of the ecosystem, one must first know elephants and know them intimately, thoroughly and completely.

An opportunity to do this came when two orphaned young elephants were brought in, abandoned at the height of the dry season near Aruba dam, when they were too weak to follow the herd back to the feeding grounds many hot, dusty miles away. It was indeed opportune that fate had made them available at a time when they were needed most, for they would lead to a better understanding of the part played by their wild counterparts in the extremely complex ecology of the Park. And from that moment on, at no time has Tsavo East been without orphaned tame elephants. Those first two, Samson and Fatuma, have been followed by many others over the years, who come in need of care, and who go when they wish, but all of whom have added their contribution towards a knowledge, understanding and a love of elephants.

The newly trained Field Force returned from their first patrol having made a marked impression on the poaching fraternity. They had shot a poacher dead when he had fired a poisoned arrow at the leading Ranger.

The Force of thirty men was divided into three Sections, with a Corporal and Lance Corporal in charge of each. In overall command was an ex-army Somali Sergeant with many years' soldiering behind him. The Sections were all equipped with pack radio sets, and were controlled and directed from headquarters by radio, so that leakage of information concerning their proposed movements was impossible. The discipline in the Force was very strictly enforced, and David insisted on instant and unquestioned obedience of any order given, whatever the difficulties involved. An opportunity to put this to the test presented itself one day in the form of a crocodile.

While travelling along the road bordering the Galana River, David noticed a large crocodile lying on a sandbank at the edge of the water. He stopped his car, and taking cover behind some bushes, crept up to within a few yards of the reptile to take a photograph. Having done this, he stood up, expecting the crocodile to slither into the river, and was surprised when it continued to lie there with its eyes closed. He then tossed a pebble at it, wondering if it was dead, but it slowly opened one eye, remaining where it was. A second pebble brought a slightly more definite reaction – it opened and closed its mouth. David soon realized that the crocodile was unable to do more than this, being completely paralysed, although no sign of any injury was visible. He returned to his car, marked the place by leaving a broken branch across the road, and drove back to headquarters where he sent for one of the Field Force Corporals and gave him the following order:

'Take your Section down to the river, and follow the road to Sobo, until you come to a place where I have put a green branch on the road. At this point you will leave the vehicle and walk down to the river, where you will see a crocodile. You will catch this crocodile, and bring it back to headquarters alive. Is that understood?'

'Yes sir,' replied the Corporal with a hint of astonishment. 'How do you want me to catch it?'

'That's your problem! I suggest you take a rope,' said David.

'Very good, sir,' replied the puzzled Ranger, taking a smart step backwards to salute. David was amused to hear him call up his men and tell them they were going down to catch a crocodile and bring it back to headquarters alive. 'A crocodile!' echoed one of the Rangers incredulously, whereupon the Corporal swung on him angrily and snapped, 'What do you think I said!' There followed a deathly hush as the Rangers scrambled into the truck without further questions, and drove off looking rather puzzled.

Three hours later, they were back, triumphantly singing on the back of the lorry, with the unfortunate crocodile trussed up like a chicken. The Corporal reported to David, saying as he saluted, 'The crocodile you ordered me to fetch, Sir.'

'Good. I want it skinned carefully so that I can see what is wrong with it.' But David never did find out the reason for the crocodile's paralysis. There was no evidence of anything untoward with his spinal column.

Although the Field Force was proving very effective, the

poachers, on the other hand, were becoming more adept at evading capture. They avoided operating along the river itself and tried to waylay their victims on paths and game trails some distance from the river. When fetching water, they were careful not to leave any tracks, walking down the many rock seams converging on the river, or even wading in the river itself. On one occasion, a patrol surprised a poacher in the act of crossing a sandy *lugga*, and this man had actually gone to the lengths of tying grass sandals to his feet in an attempt to conceal his spoor.

One of the patrols operating on the north bank of the Galana near a place called Lali, came upon the footprints of six men, which they followed for three miles upstream to a point just short of Sobo. Here, one of the Rangers, who had covered this area on a previous patrol, knew of a deserted hide-out on an island in the river, and as the tracks appeared to be leading in this direction, the patrol commander decided to wait until nightfall and approach the place under cover of darkness. At about 11 p.m. they crept up towards the hide-out with the intention of surrounding it, but unfortunately their presence was detected by a poacher, who gave the alarm. The gang bolted *en masse* from the hide-out, but were intercepted by that half of the Section who were trying to work round the back. Three of the poachers then turned back, but one promptly attacked the patrol commander, who just had time to fire one round before having to grapple with his opponent. A desperate struggle followed; the two men locked together, rolling over rocks and thorn scrub, each trying to overpower the other, until, luckily, another member of the patrol rushed up and by hammering on the poacher's hand with the butt of his rifle, forced him to release the knife and managed to overpower him.

Some while later, the handcuffed poacher complained that he had been wounded, and so the Rangers investigated the complaint by the dim light of a torch and found that what the poacher said was quite correct: he had a bullet hole through the flesh of his leg! This caused some excitement, and everyone crowded around to have a look. However, they managed to placate the poacher by convincing him that he was very lucky not to have one through the head! Back in the hide-out, a heap of incriminating evidence was collected: four tusks, bows and poisoned arrows, some drying meat and stinking hide, two rhino horns and a giraffe tail, all of which sent the poacher to jail for eight months.

The broken country on the upper Tiva was a hot-bed of poaching activity, and the many small springs on the upper part of the Yatta provided ideal vantage points commanding a good view of the surrounding country, where hide-outs could be established. Because of this, it was very difficult country to patrol, but it wasn't long before the Field Force became familiar with all the game trails, hide-outs and sources of permanent water, and by laying ambushes, managed to apprehend a good number of the Wakamba culprits.

Unlike the Waliangulu, who made their hide-outs in thick bush, the Wakamba preferred the tops of rocky outcrops and hills, so that they could detect anyone approaching some way off. The Rangers therefore had to adopt tactics to counter this, and by circling any hill at a distance, and searching all round for tracks, they got a good idea of whether a hide-out was in the vicinity. If they happened to find footprints leading towards a hill, and no evidence of the poachers having left, they would wait until nightfall, surround the hill and gradually work their way to the top. The Wakamba were quick to resort to their poisoned arrows when pressed, and many Rangers, over the years, have narrowly escaped death. One extremely lucky man just happened to turn in time for the arrow to penetrate his water bottle instead of his chest.

As time went on, it became increasingly clear, that in order to put a stop to poaching within the boundaries of the Park, the offensive would have to be carried to the poachers' actual settlements within the African reserve. They would have to be pursued and harried until such time as they accepted that the repercussions of poaching in the Park outweighed the advantages. One of the principal Waliangulu poacher strongholds was a village known as Kisiki-cha-Muzungu, some twenty miles from the Park's eastern boundary. It sheltered many of

Elephants seldom fight seriously

the most active poachers, and merely consisted of an aggregation of huts dotted at random in an area of approximately four
square miles, which were connected by a spiderweb of small
footpaths.

A series of night raids was launched on this and neighbouring
Waliangulu settlements, and it was on one such raid that David
first came face to face with a man called Galogalo Kafonde, who
was almost a legend amongst the Waliangulu tribe. It was said
that he had killed literally hundreds of elephants, many of them
large tuskers, and he was regarded as one of the most fearless
hunters, highly skilled in bushcraft and the art of avoiding
capture. Four of his sons were also well-known and appeared
to be following in his footsteps. Galogalo was reputed to know
practically every elephant path and water-hole in the Park, and
he featured high on the list of those most wanted for poaching
offences.

The raid was planned for dawn, so the anti-poaching patrol,
led by David himself, drove as far as possible under cover of
darkness, left the vehicles in the bush when some miles from
the settlement, and marched the remaining distance. As they
approached the first few huts, they heard some whistling and
humming nearby, so they crept stealthily up and found a man
clinging to the trunk of a coconut palm, busy collecting the sap
in a gourd. From this liquid, the Waliangulu brewed their
famous palm wine, and it is at night, when the sap in the tree
is rising, that they tap the palm. David ordered the sap-collector
down in a hushed voice, warning him not to make any sound.
The astonished man literally dropped out of the tree and landed
in a huddle at its base, where he was hurriedly seized. After
questioning him about the locality of the various huts and to
whom each one belonged, he was firmly handcuffed to a Ranger,
and told to lead the way.

By this time, dawn was just beginning to light the sky, and
the birds were starting to wake. The first hut was quietly
surrounded, with two Rangers guarding the entrance, while
David knocked on the door and went inside to search. Pieces of
meat, bows and arrows, some tusks tucked under a bed, and

several elephant tails were unearthed, so the menfolk were handcuffed and held for further questioning, while the patrol concentrated on the next hut, and so on, until there remained only one group at the far end of the settlement. It was here that they came upon Galogalo seated just outside the door, quietly watching the dawn appear. If he was surprised at the approach of the patrol, led by a European, he didn't betray it, but remained where he was, making no attempt to escape, and rising as David walked up to him and said,

'What is your name?'

He hesitated before replying. 'I am Galogalo Kafonde.'

'I wish to search your hut,' said David trying to hide the excitement he felt at this unexpected success.

'You may do so,' Galogalo replied, with dignity.

It wasn't long before the search bore fruit, and out of the hut came several skins, some rhino meat and a huge bow, plus various other paraphernalia, but no ivory. Men were detailed to scour the bush, and sure enough, hidden in some undergrowth were two large tusks. Galogalo was handcuffed and taken to join the other prisoners.

Then began the tedious business of questioning all the suspects, which took until 3 p.m. Whose hut is this? Then these tusks belong to you? Who shot the elephant then? Who was with you at the time? Where? When, etc., so that brief statements could be recorded. Any Waliangulu against whom there was conclusive evidence, were made to stand to one side, and the others were finally released. When Galogalo was interrogated, he admitted to having killed the elephant whose tusks had been found outside his house, and said that he had shot it while on safari in the Park with his four sons, who were still operating.

'Do you know the hide-out your sons are using?' enquired David.

'Of course,' was the reply.

'Then you will take us there.'

'Let us go,' he said simply.

They set off with a small group of Rangers, leaving the rest of the patrol behind to guard the prisoners and trophies until their return. It was just getting dark when they left the vehicle at a point some twenty miles inside the Park to proceed on foot towards the Yatta Plateau, where the hide-out was reputed to be. Galogalo led the way, and with a length of cord secured from his handcuffs to the Ranger behind, they marched through the bush for about two hours. Suddenly Galogalo lurched forward, tugging the cord as he did so. The Ranger at the other end automatically pulled back just as Galogalo let go of his end, and promptly fell flat on his back, while Galogalo crashed off into the darkness. Obviously he had managed to slip his hands out of the handcuffs fairly early on, and had been merely holding them, awaiting a suitable opportunity to make good his escape. It was hopeless trying to follow, for by the time everyone collected their wits again, he had too great a start.

The disappointment following this incident was offset to some extent by the news of the arrest of a most notorious Wakamba poacher; a man named Wambua Makula, who, for many years had associated and hunted with the Waliangulu tribe in preference to his own. This represented a major breakthrough in the anti-poaching effort, because Wambua was probably even more renowned than Galogalo. Information concerning his whereabouts was volunteered unexpectedly by a Waliangulu poacher arrested for being in possession of ivory during a routine patrol. He disclosed the fact that Wambua was currently living near the coast at a place some thirty miles from Kilifi. This news was hastily relayed to the Game Department at Kilifi, but Wambua had, in fact, already been arrested by them some two hours earlier following the same information from a different source. He was then despatched under escort to Voi to answer the formidable list of alleged misdemeanours which appeared on his card in the Rogue's Gallery.

The source of the information which led to his capture was never disclosed to Wambua, but nevertheless, he was convinced that it had been the Waliangulu who had betrayed him and led to his downfall. Probably professional jealousy motivated this belief, but with every day that passed, he became more determined than ever to get his revenge and thus even up the score. This was extremely fortunate for he poured out a wealth of valuable information about many wanted people, and answered any question in great detail. What he didn't know about the activities of most people was no one's business, and the Rogue's Gallery swelled considerably. Wambua proved so co-operative that the charges against him were waived, and he was instead attached to the staff as official informer cum interpreter.

In the weeks that followed he led the patrols to all the places most frequented by poachers, pointing out the hide-outs as well as three hitherto unknown secret sources of permanent water on the Yatta, used by poachers to replenish their supplies when the river was heavily patrolled. Slabs of rock carefully concealed three deep holes in the lava which retained water for many months after the rains. At these places, a grim reminder of the extent of the poaching was seen in the carcases of five rhino found nearby.

The Field Force, after they had been operating for one year, had gone a long way towards controlling the poaching in the Park itself, particularly in the southern section. The northern area, where communications were still difficult, was more of a problem and poaching there was still rife.

Morale amongst the Rangers was extremely high, and the discipline tight; an essential in view of the fact that they were all armed. Very strict orders regarding the use of firearms were enforced. Only in self-defence, or in defence of a fellow Ranger in grave and imminent danger, was a bullet supposed to be fired. When, for some reason, a firearm had been discharged, there followed a full enquiry into the circumstances. This imposed a considerable test of discipline on the Rangers, for they could not always be certain of a poacher's intentions, and knew only too well the consequences of being struck by an arrow, the poison of which had no known antidote. In time a force was built up of men whose courage had been put to the test many times, not only when tackling poachers, but also when confronted by dangerous animals. They were proud of being able to overcome hardships during the course of their duties, and were capable of great feats of endurance when necessary. Each man knew that he could rely on his fellow Ranger, and in this way a tremendous *esprit de corps* was fostered throughout the Force. Very little was known of the Field Force outside the National Parks' organization; no medals were handed out for bravery, nor were their praises sung by a grateful Government. Their knowledge of bushcraft, tracking and general ability in the bush rivalled that of the poachers themselves, and this, coupled with the iron discipline imposed on them at all times, produced a formidable unit whose reputation slowly spread far and wide, and earned the deep respect of the poaching fraternity.

Having considerably cleaned up the area south of the river, the drive against the poaching was carried into the northern area, when with the completion of the causeway, work was put under way to open up this vast tract. A start was made in this direction by cutting a road from Lugard's Falls through the heart of the northern area to the Ithumba massif on the northern boundary, a distance of seventy-one miles through desolate, waterless country. When the road reached a small hill resembling a pyramid, which rose abruptly from a dead flat wall of bush, and was known as Kiasa, a 6,000 gallon concrete water tank was constructed below the rock face in order to conserve the run-off during the rains for the use of patrols operating in the area east of the Yatta escarpment and between the Tiva and Galana Rivers. And once this had been completed, the Field Force became very active.

One patrol intercepted the spoor of two men, which they trailed over parched gravelly terrain for a distance of ten miles, before suddenly detecting the smell of wood smoke which could only mean that they were in the vicinity of a hide-out. They spread out and moved forward stealthily, and had gone a short distance only when one of the Rangers heard a slight rustling sound to his left. He swung round, and an arrow hit the ground at his feet. Hurriedly looking up, he saw a man in the process of fitting another arrow to his bow, so instinctively he raised his rifle and fired. As he rushed forward, a second poacher leapt from the bush and seized hold of his gun, attempting to wrest it from him. The two men struggled for possession of the rifle until the remainder of the patrol came to the rescue and succeeded in overpowering the poacher, binding his wrists together with their whistle lanyards. Everyone was so preoccupied with trussing up their prisoner, that they had forgotten all about the presence of the first poacher, until a faint groan drew their attention. To their surprise there was the man lying on the ground, writhing in agony. A closer examination revealed why – the bullet had shattered his right shoulder and lodged in his stomach.

Carefully lifting him, they carried him gently to the shade of a nearby tree, and there did their best to staunch the flow of blood with strips torn from their clothing, but alas, to no avail. The poacher died shortly afterwards.

Detailing two Rangers to guard the body, the Patrol Commander returned to camp with the one prisoner and the rest of the patrol, where he immediately contacted H.Q. by radio to report what had taken place. The Ranger was vindicated following a thorough police investigation at the scene of the encounter. The verdict was that he had been justified in firing in self-defence, and had acted in a responsible manner.

News of this incident must have rapidly travelled on the 'bush telephone' to other poachers operating in the northern area, because subsequent patrols came across a large number of recently vacated hide-outs and signs indicative of a hurried departure. One poaching gang, however, was encountered, and it was a rhino they had to thank for their escape. Just as the Rangers were creeping up on the hide-out, a rhino exploded out of a thick clump of bush and hooked one of the crouching Rangers between his bushjacket and shorts, tearing the jacket clean off his back. Very fortunately, however, the rhino did not persist and carried on its way, while the Ranger dived head first into the nearest bush! Poachers streamed out of the hide-out, and the diversion caused by the appearance of the rhino enabled them to make good their escape. There was another similar episode, but on this occasion it was a herd of buffalo that stampeded through the midst of a patrol just as they were in the process of surrounding a hide-out.

While most of the anti-poaching activity was now concentrated in the northern area, it was equally important that the initiative which had been gained south of the river, was not lost, so the pressure was kept up by continued night raids on the various Waliangulu settlements. Most Waliangulu poachers now avoided hunting in the Park and only the 'die-hards' persisted.

Quite obviously, it would not be possible for the Field Force to maintain control indefinitely over the adjoining areas, as well as the Park itself, but David realized that any relaxation at this stage would only encourage the poachers to return to their old stamping grounds. An approach was therefore made to the East African Wildlife Society with a plan aimed at combating poaching in the entire Coast Province and part of Central Province, and this was eventually submitted to the Government. It called for a concerted all-out anti-poaching effort and recommended the establishment of two additional Field Forces in co-operation with the Game Department, one to be based at a place called Makindu to deal with the Wakamba poachers along the Park's northern and western boundaries, and the other to operate from Hola on the Tana River, and work back towards the north-eastern boundary down the Tana to the coast. The existing Field Force would be known as Voi Force for the duration of the campaign, and would continue to be responsible for the Park itself and the areas south and east as far as the coast.

The plan was discussed at a meeting of interested parties which included representatives from the Game Department, National Parks, the police, the army and the Wildlife Society, and agreement was finally reached on the best method of conducting the campaign. The army agreed to release sufficient radio equipment to provide complete radio coverage, while the police seconded two officers, one to head investigations into the traffic in illicit trophies, and the other to assist with prosecutions.

David was asked to command the campaign and the Game Department agreed to release several officers who would be in charge of the two new Field Forces which were to be recruited and trained at Voi without delay. Until such time as these two new Forces were operational, the existing Field Force were to redouble their efforts in order to hold the initiative both inside and outside the Park, even though this meant operating over an extremely extended area.

Although the hunting blocks around the periphery of the Park were regularly covered by professional hunters and their clients, this did not deter the poachers operating in these areas, and, indeed, some were even brazen in their disregard. On one occasion, a well known professional hunter, accompanied by an American client, had actually drawn a bead on a large bull elephant near Mackinnon Road, when two arrows whistled past his ear in quick succession, and lodged in the animal he was just about to shoot. The elephant trumpeted and tore off into the bush, hotly pursued by two Waliangulu poachers, who sprinted past the flabbergasted hunter and his client without even a sideways glance. Understandably, the hunter was greatly incensed at being robbed of a good trophy, and hastened to the nearest railway siding to despatch a telegram to the National Parks.

The Field Force were there within an hour, and picked up the tracks of the two poachers without any difficulty, which they followed until nightfall. Sleeping on the spoor, they continued the chase at dawn the following morning. After a few hours, it was noticed that the tracks diverted from those of the elephant they were following, and trailed the spoor of another elephant, which was found dead some five hundred yards from this point. There was, however, no sign of the poachers, but the tusks were still intact, although some meat had been taken from the carcase, and it was obvious that the poachers would be returning to collect their loot. Hurriedly the Rangers concealed themselves around the carcase, and they did not have long to wait, for very soon the two poachers came striding jauntily along, laughing and talking in loud voices. The Rangers waited until they had reached the elephant before descending on them from all sides. Although the arrest of the poachers helped placate the hunter, and his client, the memory of the incident continued to make him feel sore for a long time afterwards!

As soon as the road through the northern area from Lugard's Falls to Ithumba Hill had been completed, work began on a northern Sub-headquarters, sited on the slopes of Ithumba itself. It was envisaged that an Assistant Warden would be based permanently there to ensure more effective control of the northern area. Tall, shady trees and a small spring nearby which provided water as clear as crystal went a long way towards making the new headquarters a most attractive unit. A charming open bungalow type house, a small office, workshop and garage, guard room and staff lines were constructed. The headquarters commanded a spectacular view towards Kime-thena Hill which marked the Park's northern limit. Giant baobabs towering above the thick bush traced a lacey pattern against the skyline. The atmosphere of peace was enhanced further by a natural background music of bird songs, the stirring sound of a francolin's call, the fragrant perfume of wild flowers and lush vegetation that wafted on pure, un-polluted air, and the myriads of brightly painted butterflies which adorned the scene by darting like jewels from place to place.

David undertook several prolonged foot safaris in the area, and discovered that in the Tiva River area was a unique heri-tage that would need careful guarding; it harboured an astonish-ing number of really massive tuskers; huge bull elephants that proudly carried ivory weighing well over 100 lbs. per tusk, and some so enormous that even the most experienced hunter would hesitate to hazard an estimate of their weight. The northern area was truly a land of giants; giant elephants and giant baobabs, and it held a peacefully remote fascination which could only be described as thrilling. David resolved to try and keep it this way as far as he was able, so that in years to come when progress had destroyed the atmosphere of so many wild places, people would still be able to find it here.

Chosen for the post of Assistant Warden Northern Area was my brother, Peter, who was to be transferred from Tsavo West to take up his new appointment, and supervise the finishing touches to the new headquarters.

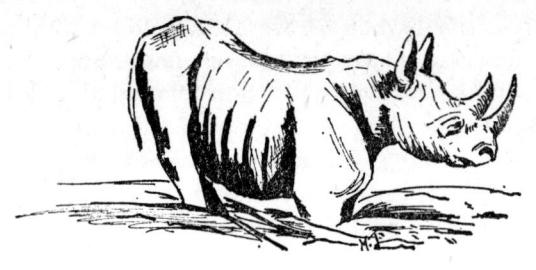

CHAPTER 6

The Anti-Poaching Campaign

BILL served with the Kenya Regiment for a period of three years, and during this time we had married, and our daughter, Jill, had been born in Nairobi. Bill had served with distinction, not only attaining the rank of Captain, but being awarded the Military Cross for bravery in the forest. The army had made a man of him, and he left it a far more mature and responsible person.

We looked forward eagerly to being able to settle into our new house in Voi, and lead a more peaceful existence, but at the same time, I harboured secret misgivings about the prospect of living there, for the rather adverse impression gained during my one visit to Tsavo East lingered. In addition to this Voi had the reputation of being an extremely unhealthy district with a debilitating climate rife with malaria. Fresh provisions were difficult to procure and the responsibility of motherhood weighed heavily on my inexperienced shoulders, for I was only twenty at the time. I wondered how my small daughter would get on with the scorpions, bird-eating spiders, venomous snakes and numerous other hazards she was sure to encounter when she reached the crawling stage. But, I kept these anxieties strictly to myself, not wishing to tarnish my image as a safari going member of an old pioneering family!

The day arrived when we were ready to set off for Voi. Bill followed in a lorry loaded with all our possessions, while I went on ahead in the car with Jill, but I found that I had some explaining to do when we met up at the other end, for I succeeded in colliding with a donkey *en route*. It was difficult to explain to Bill that the donkey charged straight at the car from the side of the road and that the damage had been no fault of mine.

The weeks that followed our arrival in Voi saw all my fears slowly subside as I found myself totally absorbed by the task of settling into our new house, planning and planting a garden, making curtains and painting and polishing until everything was to my liking.

Meanwhile the personnel for the anti-poaching drive were busy collecting at Voi and the recruits for the two new Field Forces were undergoing training. Seconded from the Game Department for the duration of the campaign were three European officers; David McCabe, Dennis Kearney and Ian Parker, as well as a fourth loaned by the Administration, David Brown. Major Hugh Massey also joined the anti-poaching effort.

David had at his disposal a police Airwing plane and pilot, plus access to the Kenya Police radio network throughout the country, while the man chosen to head the team which would be concerned with apprehending the ivory dealers, and disrupting the traffic in illicit trophies, was Senior Superintendent Rossie Potgieter, who would be based at Mombasa and liaise closely with Voi.

While final preparations were made, a host of informers were employed to collect details about the activity of poachers and buyers, which would supplement that already recorded. Many conflicting reports were at first received, but slowly a complete picture began to emerge, and the stage was set for the onslaught.

The day arrived when the new Field Forces were ready to become operational, and the officers were allocated their positions. David McCabe and David Brown were detailed to command the one to be known as Hola Force, while Major

Hugh Massey, Dennis Kearney and Ian Parker were attached to the other which was to be based at Makindu in the Central Province, and would be known as Makindu Force. The Parks Field Force, which would hitherto become Voi Force, became the responsibility of Bill and my brother Peter, while David would direct all operations from headquarters, now given the title of Poacher Control.

The campaign started with a flourish and a series of night raids on various settlements around the periphery of the Park. It was clear that all the painstaking ground work was now bearing fruit, for news poured into headquarters at each radio call, that many wanted men had been arrested. In addition, a great number of trophies were recovered, and weapons, traps and poison seized. The three Field Forces vied with each other for the greatest success and lorry loads of prisoners and their tackle returned to headquarters for interrogation and prosecution. There was no doubt that the campaign, although only in the initial stages, had already gone a long way towards disrupting the whole traffic in illicit trophies. Dealers no longer knew whom they could trust, while poachers went in constant fear of being arrested.

David conducted the campaign with cool and calculating precision, and this, coupled with the individual ability and leadership qualities of the Force Commanders themselves, welded the anti-poaching Forces into a formidable unit.

The main bottleneck lay in correlating the mass of information that kept pouring in, and in recording statements from prisoners and preparing all the Charge Sheets. Finally, I was asked if I would help with this work and in the months that followed I, too, found myself deeply involved in the anti-poaching effort, the now much enlarged Rogue's Gallery being my particular responsibility, while Jill happily played outside the office, scrambling amidst a heap of tusks and other poaching paraphernalia under the eagle eye of an African nanny engaged to see that she didn't wander, or pick up those scorpions, bird-eating spiders and venomous snakes which were always at the back of my mind.

The intelligence system that was built up was extremely efficient and effective. Even the family relationships between members of the Waliangulu tribe became known and was recorded on an impressive 'family tree', fitting into place like the pieces of a jigsaw.

The actual interrogation of prisoners fascinated me, for there was a definite art in getting them to talk, simply by subtle insinuations and roundabout questions, for at no time were the crude old-fashioned methods of extracting information allowed. For instance, during the course of questioning, a poacher might reveal that his colleague had turned back because he had been stung by a scorpion on his toe while trailing an elephant, and even this seemingly irrelevant information would be recorded on the appropriate card. Finally, when the companion was brought to book, he was flabbergasted by only one question, 'How's your toe? Have you recovered fully from the scorpion sting you suffered?' Invariably, the poacher would be so dumbfounded by this that he very often assumed logically that if this small incident in his life was known, what was the use of beating about the bush, because everything else must be known as well! More often than not, he proceeded to pour out a string of confessions relating to all his misdeeds, and if he was a little reluctant to talk, there was always Wambua, standing there like his conscience ready to prompt his memory.

On many occasions, Samson, the tame elephant, unwittingly loosened tongues, for he had a habit of strolling around the offices looking for David, and inevitably he had to pass a long line of squatting prisoners, who were awaiting their turn to be questioned. Sometimes he would extend his trunk to sniff one of the poachers, in all probability merely hoping for a hand-out, but Wambua, who was adept at seizing the advantage presented by this simple action, would exclaim triumphantly,

'Ha! You see, the elephant knows you killed his mother!'

This never failed to unnerve the poacher completely, for, steeped in witchcraft and superstition, he attached a hidden interpretation, particularly as an elephant was involved. It would prey on his mind until he was only too happy to attempt to appease the ghost of Samson's mother by making a clean breast of everything!

But, in spite of all the other successes of the campaign, the principal prize, in the person of Galogalo Kafonde, continued to elude capture. However, his activities led instead to the arrest of a notorious dealer.

Galogalo, with a party of other Waliangulu, had shot a number of elephant near Lali. They recovered the tusks and carefully sawed off the unwanted butt ends, tossing them into the bush. The ivory was then sold to an African middleman, called Kisau, who engaged porters to carry it to the coast for onward transmission to the Asian dealer.

Meanwhile, one of the Waliangulu who had accompanied Galogalo on this hunt, happened to be arrested by Voi Force in a surprise raid on Kisiki-cha-Muzungu, and it was he who revealed what had taken place. Bill quickly passed the information by radio to David, who lost no time in setting out to look for Kisau. Fortunately, he managed to unearth a youth in Voi who happened to know where Kisau lived, and who volunteered to lead him to the place.

They drove, under cover of darkness, to a point about thirty miles short of Mombasa, and left the car by the side of the main road to walk the rest of the way on foot. It was about 2 a.m. before they finally emerged into a small clearing in the bush, and could see a group of huts surrounded by cultivation in the faint light of the moon. They approached cautiously and a Ranger was quietly posted to guard the entrance to each hut, before the search began.

While David was occupied searching the first hut, there was a crashing sound and a shout, as a naked man burst out of one of the other huts, knocking over the huge Turkana Ranger guarding the entrance in the process. The Ranger was up in a flash and chased after the man like an enraged buffalo. Shortly afterwards, yells and a sound like someone beating a carpet could be heard coming from the bushes. Faintly alarmed by the prospect of what was taking place, David hurried to the spot, where he found the infuriated Ranger busy belabouring a man

79

who was lying on the ground doing his best to ward off the hail of blows. The Ranger, who was incensed beyond all proportion at the indignity which he had suffered, took quite a lot of pacifying, and seemed to be bent on nothing short of murder for revenge.

Before the man on the ground could recover his wits, David was firing questions at him,

'Kisau, where is the ivory you bought from Galogalo two days ago?'

'I have already sold it,' he blurted out.

'Who to?' David demanded.

'To an Asian in Mombasa,' he replied. At this point he suddenly seemed to realize the significance of the questions, for he was reluctant to answer the next.

'Where is the money you received?'

Kisau shuffled his feet and glanced nervously up at the Ranger towering over him, who took a menacing step forward. The sight must have unnerved him, for he then hastily answered the question,

'I ran out of the hut with the money, but dropped it when I was caught.'

There followed a search by torchlight, but the money could not be found.

'Are you sure you are telling the truth?' snapped David. The Rangers let out a low growl, making it plain that they were sure he wasn't!

'Yes, yes; I swear,' stammered Kisau desperately, almost in tears.

'All right, then we will stay here until dawn and find the money in daylight,' David announced.

During the long wait until dawn, Kisau seemed to become reconciled to making a clean breast of everything; and in fact appeared almost eager to do so. It transpired that he had sold the ivory to the Asian, but had not yet been paid the full price, and had intended to return to Mombasa to collect the balance in two days' time. Here, clearly, was just the opening everyone was looking for! David explained to Kisau that while he would

A typical Tsavo sunset

be charged for the offence he had committed, any assistance he might be prepared to give to the authorities, would be brought before the notice of the Court at his trial, and would prove to his advantage, so Kisau readily agreed to help.

As soon as there was sufficient daylight, the search for the money continued, and sure enough, after only a few minutes, there was a shout of triumph as one of the Rangers came running up with a wallet full of notes.

The next step was to contact Rossie Potgieter in Mombasa and decide how to proceed. They had a quick, discreet look at the Asian's premises, and decided that Kisau should keep his appointment with the Asian, and that a police constable in plain clothing should be stationed in the narrow street outside the building where he could see what was happening through the open doorway.

The important day arrived, and everything worked according to plan. Kisau kept the appointment, was paid the money owed him and the constable, mingling with the crowd outside in the street, actually saw the money being handed over, although he could not hear what was being said.

So far, so good, but Kisau was the bearer of some disturbing news when he reported back to Rossie and David. When he asked the Asian what he had done with the tusks, the Asian informed him that they had already been moved to another godown and that work on them had started. This meant that there was not a moment to lose. David and a police constable watched the building Kisau had just left, in case the tusks were, in fact, still there, and might be moved somewhere else, while Rossie and Kisau rushed round to the godown, which they found locked. There was no response to a knock on the door, so Rossie clambered up the wall to peer through a small grille. There on the floor sat an African, frantically rasping and adzing some pieces of ivory. Again there was no response when Rossie demanded that the door be opened in the name of the police, and the man merely worked even faster at what he was doing. In desperation Rossie hammered and battered the door, which attracted a large crowd as people came running to the spot, some of them even brandishing sticks, thinking that they were about to catch a thief! The commotion grew until it looked as though an ugly situation might be developing, so Kisau scuttled down the street to fetch David and the other police constable, and back they all ran to Rossie's aid, expecting to find him lynched! They arrived instead to find that he had the situation well under control, having successfully convinced the crowd that he was in fact a police officer. Someone had then gone to call one of the owners of the godown, who rather reluctantly opened the door.

The place was stacked with ivory of all shapes and sizes. Although most of it had been purchased legally at the Government Auction, when the books were scrutinized it soon became apparent that a good deal could not be accounted for, and the owner could offer no satisfactory explanation regarding the discrepancies. Kisau examined the pieces of ivory that were in the process of being rasped, and was able to identify some of the chunks that had yet to be adzed, as those he had sold that morning. The godown was therefore locked and placed under police guard and the owner informed that he would be charged with being in possession of illegal ivory.

The case against him, however, was far from watertight, for the only evidence was that of an accomplice. David then had a brainwave – somehow the tusk butts which were lying in the bush must be retrieved and matched to the pieces in the godown, so that there could be no doubt that the ivory was, in fact, illegal.

This was explained to Bill over the radio, who immediately set off on foot with the Waliangulu informant to search for the butts, while Rossie and David anxiously awaited the outcome so that they could prepare their case. Two days later word was received that the butts had been recovered, and were already on their way to Mombasa. On arrival they were fitted to the pieces in the godown and the matter was clinched, for every hair crack corresponded. There was no doubt that the pieces in the godown, and the butts which had been lying over 100 miles away in the bush, were part of the same pair of tusks. And so the accused didn't have a leg to stand on and was duly convicted and sentenced to a term of imprisonment.

Things did not always go as smoothly as this, however, for the ivory dealers were extremely slippery customers and could afford to sacrifice huge sums of money in order to hire the best lawyers to defend their case. They were not above bribery, either. The prosecution, therefore, had to be very sure that all loose ends were securely buttoned up in order to obtain a conviction. A clear example of this was the occasion when an Asian dealer was actually caught red-handed loading illegal ivory into his car, and the Game Department Officer had snatched the ignition key from the vehicle as a precaution against an attempted get-away. When the case was heard, the defence council submitted that it was not the Asian who had possession of the ivory, but the Game Warden himself, for was it not he who had the ignition key and hence control of the contents of the car! Fortunately, the Magistrate dismissed this argument, and the dealer was, in fact, convicted.

Rossie Potgieter's role in the campaign was an unenviable one, although at times it must have proved extremely exciting. He was subjected to all the intrigues of the underworld in Mombasa's Old Town, and his investigations took him into the most extraordinary and sordid surroundings. He made the acquaintance of a host of unsavoury characters who looked as though they might slit your throat as soon as look at you! And he found also, that the more he uncovered, the more snarled up and complicated things became, so that the repercussions of any action he tried to take were often far reaching and unexpected, which made his job doubly difficult, if not almost impossible.

There were times when all three Forces happened to be back at base together, and such occasions always called for a celebration at the Voi Hotel, where everyone forgathered to relax and compare notes. Dave McCabe was the life and soul of any party, being an extremely witty character, whose unpredictable actions guaranteed a fund of amusing stories that provided many lighter moments. Ian Parker was another born comedian and good for an evening's entertainment. He greatly enjoyed getting a rise out of Hugh Massey, who was apt to give the impression of being slightly disapproving of these wild parties, and this only made him more vulnerable.

There was one occasion when he had disassociated himself from the riotous antics in the bar, and was bending down in the adjoining lounge perusing a copy of *The Field*. The temptation of this sight proved too much for David, who plucked a dart from the board, took careful aim and sent it hurtling through a connecting window to lodge where intended; in poor Hugh's read end! Slowly and stiffly he straightened up, extracted the dart with a quick pull, swung round and puce with rage, strode purposefully into the bar. By this time, David had engaged his neighbour in rapt conversation, but Ian, who had witnessed the incident, couldn't hide the amusement he felt, and was caught by Hugh grinning with a look of satisfaction on his face. Convinced that Ian was the culprit, and justifiably even more incensed that his subordinate should have the temerity to do such a thing, Hugh deliberately walked up to him, and jabbed the dart deep into his buttocks several times, in spite of Ian's anguished protestations of innocence! These were merely drowned by cheers from the bar contingent!

The long-suffering manager of the Voi Hotel deserves special mention, for not only was Henry Hayes a typically jovial Yorkshireman and a tremendous character, but he was also a great entertainer and full of fun himself. He was an enormous man weighing close on 300 lbs. and had a red, merry face, twinkling blue eyes and a shiny bald head. He also had a magnificent singing voice, so powerful that the glasses on the bar shelves rattled when he hit the high notes, and all extraneous noises, in the bar, and elsewhere, were drowned as Henry burst into song.

One night, or to be more accurate, early morning, when Henry was in top form, a guest who happened to be staying in the hotel, and who was unfortunate enough to have been allotted the room nearest the bar, poked his head around the door, clad only in his pyjamas, and announced that some people were trying to sleep. There was a deathly hush, as everyone at the bar swung round on their stools, to stare at him in amazement and disbelief.

'Get that bum out of here!' someone muttered thickly, and with that there was a mad rush like the charge of the light brigade towards the unfortunate guest. He lost no time at all and sprinted to his room, hotly pursued by all and sundry. He only just made it, and somehow managed to round the corner at top speed, his legs scrabbling to keep their foothold as he dashed into his room and slammed the door in the nick of time. There was dead silence from within as his pursuers elaborated on what his fate would be if he appeared again.

Back in the bar, Henry had a twinge of conscience.

'Gentlemen! Please! I have sixteen guests . . .'

With that a chorus went up, made all the more deafening by the hammering of fists on the counter. 'Sixteen guests! And what do you get? Another day older and deeper in debt!'

Henry did his best to restore order, but finally was defeated and took himself off to bed. I often wonder if those sixteen guests ever patronized the Voi Hotel again.

It had taken a little time to indoctrinate Henry into accepting the rather eccentric habits of colonials, for when he had first taken up his appointment, newly out from Yorkshire, he had tried to enforce licensing hours and orderly behaviour in the bar. Everyone had been genuinely astonished when he pounded the bar counter and shouted,

'Time, gentlemen!'

'You must be joking!' exclaimed David.

A heated argument followed, in which it was the rest versus Henry, and when things appeared to have reached deadlock, Henry thought he would resort to his trump card.

'I'll call the police,' he announced defiantly.

'Go ahead,' laughed David, indicating a bleary-eyed individual slouching at the far end of the bar.

'What sort of a bloody country's this!' yelled Henry, as he stormed out. 'No wonder we're trying to get shot of it!'

This, being a rather delicate topic, triggered off a rowdy response that continued even after Henry had made a timely

exit, and the party went on regardless, although Henry thought
he had settled matters by locking the bar and the two fridges
where the beer was kept. Unbeknownst to him, David had long
since discovered a means of opening the fridge doors even
though they were locked, and the next day, when Henry came
to check his bar stocks, he was extremely surprised to find the
fridges empty of beer, but stuffed full of signed chits and notes
in payment thereof. Henry had to concede that the bar had,
indeed, done a brisk trade the previous night, and so from
that day on, if he felt like retiring during a party, he always
left the fridges well stocked so that everyone could help themselves.

There was one aspect connected with those Saturday night
socials that I learnt the hard way, and that was to be independent of the men for transport, so that I too could withdraw when
I felt like it. I found out that it was fatal to announce that one
wanted to go home, for this merely made the men more
determined than ever not to comply for fear of appearing to be
dominated by the fair sex. I discovered this at one of the very
first functions I had attended, for when I made this announcement in the early hours of the morning, Bill pretended not to
hear, David looked amused, and the rest protested strongly.

'All right then! I'm going to walk,' I declared rather un-
wisely, marching out in high dudgeon.

In the carpark I glanced furtively behind fully expecting to
see some gallant gentleman who would be prepared to sacrifice
his vanity rather than risk me walking home in big game infested
country, but, alarmingly, there was no one, which placed me in
a bit of a predicament. Determined not to risk the indignity of
having to accept defeat, I stalked on down the road, but the
further I went, the darker it became, and the slower was my
pace, until I found myself almost marking time! The shadows
cast on the road verges by the feeble light of a half moon took on
menacing shapes, and when a twig snapped in the bush nearby
I leapt involuntarily and my heart hammered so loudly that I
could hardly hear myself think. I wished fervently that I hadn't
been so dogmatic, and decided that the last thing I wanted to

do was to walk home! So I sat down on a rock by the side of the road thinking that surely the men would come at any moment to rescue me. But, I had a long wait, and as the minutes ticked by, the workings of my imagination terrified me even further, until I was firmly convinced that I was being stalked by a man-eater. I was about to make my peace with God, when to my immense relief, I heard the sound of an approaching vehicle and caught a glimpse of the headlights through the trees. Clutching desperately to the remnants of my pride, I continued down the road in what I hoped looked a determined way, until the vehicle drew alongside and stopped.

'Feel like a lift?,' someone enquired maddeningly.

'You rotten swines!' I retorted hotly. 'It would have served you right if I had been eaten!'

'We thought you were in the Ladies!' said David, and I could sense the amusement on all their faces. It would have been marvellous to have had the courage to decline the offer of a lift, and it did cross my mind, but I thought better of it, and clambered in and plonked myself down in a manner that I hoped conveyed the rage I felt.

When we arrived at the house, I was subjected to the beginnings of a lecture from my brother about how foolish I had been and how badly I had behaved.

'Oh, shut up!' I snapped, and derived some satisfaction from slamming the door in his face!

The anti-poaching campaign took nearly two years to eradicate poaching almost completely from the Coast and Central Provinces, but once this had been achieved, the Forces embarked on mopping up operations. It had come to light that large scale poaching had taken place in the past at a place called Ushingu, northeast of the Park boundary, so Voi Force moved into the area to have a look. No fresh signs of poaching were evident, but it was obvious that a tremendous slaughter of elephant had taken place there at one time, for altogether no less than fifty old hide-outs were found, plus the carcases of 381 elephants, many still marked by bunches of grass tied to poles.

A total of 92 tusks was recovered, the proceeds of which would help offset the expense of the campaign.

As a result of this successful exercise, it was decided to mount a similar search in the area between the Galana and Tana Rivers, where a comparable slaughter of elephants was reputed to have occurred. Here the carcases of some 900 elephants were found, and 352 tusks recovered, totalling 6,604 lbs plus a staggering 1,589 lbs of butt ends and pieces which were found strewn in the bush. These figures only too clearly indicated that the stern measures adopted to combat poaching came only just in time, and were amply justified, for had the poaching been allowed to continue on this scale, there is no doubt that the elephant and rhino would have been decimated.

The two areas referred to above are featureless for the most part, and the patrols experienced great difficulty in operating in such country. Many of the Rangers got lost and were forced to spend the night out in the bush, and on two occasions some men were lost for two days and were in a state of complete collapse when they were finally found. Added to the difficulty of the terrain was the fact that the elephant remaining seemed to be particularly aggressive following years of being harassed, and many Rangers had narrow escapes. One man, after an uncomfortably close brush with an elephant, found himself covered in saliva from the animal's trunk, which must have been literally poised above him! But, not only elephant were aggressive, for another Ranger was very nearly caught by a lion, and most of them were treed by rhino at some stage or another during the search.

In one of the hide-outs, a corpse was found, illustrating the fact that the poachers did not always escape unscathed either. Ex-poachers who had guided the patrols claimed that the body was that of a Wakamba poacher killed by a lion he had caught in a gin trap, and which had managed to free itself. Another version, whispered surreptitiously around the campfire, was that the man had been murdered by his colleagues for the ivory known to be in his possession. Whatever the gruesome story, that human corpse only contributed to the sinister atmosphere

Negotiating a flooded watercourse during the anti-poaching campaign

Carcases of poached elephants were found all along the Galana River

A poacher is captured

David interrogating prisoners

Waliangulu in a hide-out with strips of drying elephant meat

of the place, and the patrols were relieved when the day came for them to depart.

Another poacher who paid dearly for his misdeeds was a honey-hunter operating in the northern area, who, while attempting to rob a hive, was attacked by the bees and fell out of a giant baobab, breaking his ankle and dislocating his spine. Unable to move, the unfortunate man had to endure the continued onslaught of the infuriated bees as he lay completely helpless at the base of the tree, and as though this were not sufficient punishment, to his horror he noticed a bushfire, which must have been started by another honey-hunter, sweeping rapidly towards him. Somehow he managed to drag himself to a small bare patch of ground, which he reached in the nick of time, as by then the fire was raging all around him.

All night long he lay there, and it was only by great good fortune that he was discovered by a passing patrol, attracted to the place by his groans. They immediately took him to the nearest hospital by lorry, which was at least fifty miles away, and so grateful was he for their assistance, that he vowed, should he be lucky enough to survive, he would return and work free for the National Parks forever more. He never showed up, however, and so presumably he must have died, or perhaps been left so badly crippled that he could not keep his promise.

The time was now rapidly approaching when David could hand back the responsibility for anti-poaching work outside the Park boundaries to the Game Department and police, whose commitment it rightly was, and be able to recall his own Field Force to concentrate again on the Park itself. In his report to the Trustees in December 1957, he wrote:

'The anti-poaching campaign has taken up a good deal of our time, and Park's work has suffered as a result. Nevertheless, we consider it time well spent and hope that the game, particularly in the Eastern side of the Tsavo Park, will respond to our efforts by showing an increase in numbers over the next few years.' This was a prophetic statement; particularly as far as elephant were concerned!

The campaign closed on a dramatic note, with the sudden

appearance of the famous Galogalo Kafonde, who strode unexpectedly into the Malindi police station and announced that
he had come to surrender! He was tired of being constantly
harried, tired of the cat and mouse game in which he had been
involved and which had denied him a decent night's rest for so
long. He was eager to pay the penalty and start again afresh.

It was with an easy mind, therefore, that everyone returned
to their normal duties, knowing that poaching was at last under
control both inside and outside the Park. The Waliangulu,
particularly, had been taught a very salutary lesson, and
practically every male member of the tribe had seen the inside
of a jail. From that day to this, no Waliangulu has been arrested
within the Park boundary, and our elephants in the southern
section at least, have been able to live in peace.

CHAPTER 7

The Aftermath

WITH the end of the anti-poaching campaign, David started giving serious thought to the future of the Waliangulu tribe as a whole. Quite obviously, if they were to abstain from poaching, they must be afforded an alternative means of livelihood, for the number that could be absorbed by professional hunting firms as gunbearers and trackers was not very substantial, and the work was usually not on a permanent basis. Ian Parker and Noel Simon, who was Chairman of the Wildlife Society, were two others who were very concerned with this problem, and together they formulated a plan to be placed before the Government, whereby the Waliangulu as a tribe would be permitted to utilize the game in the area between the eastern Park boundary and the coast, cropping it on a sustained yield basis under close supervision, and marketing the trophies and the meat legally. The object of this scheme was to make it possible for the Waliangulu to continue to exist in the traditional style without depleting stocks of game or having to resort to illegal practices.

The Government accepted the idea in principle, but imposed a serious obstacle to the ultimate success of the scheme by insisting that while meat and hides could be marketed, any ivory or rhino horn must remain the property of the Government, and the proceeds from this source channelled back into general revenue. This, of course, deprived the scheme of its most valuable source of income from the start and had a ham-stringing effect on its viability. However, following protracted negotiations, it was agreed to implement the scheme for an initial period of trial, and see how things progressed.

The scheme was to be known as the Galana Game Management Scheme and was to be the responsibility of the Game Department. The man selected to get it under way was, fittingly, Ian Parker.

On the face of things, the idea had been a sound one, but the scheme soon foundered. This was due to several factors, one of which was the reluctance of the Waliangulu themselves to keep up a sustained effort, for they periodically drifted off back to their villages to relax and enjoy their earnings. Another reason for the failure of the scheme was a flaw in the concept of being able to crop game on a sustained yield basis and make it pay. It is one thing to ranch cattle, to be in a position to remove predation, and to keep the herds alive by inoculating them against diseases that would in the normal course of events account for a good proportion of them; one thing to be able to drive the stock to market and slaughter them at a place chosen for convenience; but quite another thing to 'farm' game and make it a paying proposition on a sustained yield basis. For one thing natural predation must be taken into account, losses are inflicted by droughts and disease; it is costly to recover the meat which might involve a great deal of travelling to reach the place where the animal was shot; the difficulty of preserving it and keeping it fresh until it can be sold, of having to transport it many miles to population centres; quite apart from the fact that the game itself, being continually harassed, becomes more and more difficult to approach, or even moves out of the area. Many game populations in Africa have been 'mined,' but few have been 'farmed' successfully and economically. If one could domesticate wild herds, inoculate them and feed them, herd them and control them in the same way as domestic stock, then perhaps it would be a different story, but as things stood at the time in Kenya, it turned out to be another of those theories that sounded promising on paper, but which did not work in practice. And so, after several years of struggling to prove the impossible, the Galana Game Management Scheme finally folded up. Ian Parker left the service of the Game Department, and established his own firm known as Wildlife Services Limited with the object of setting himself up as a consultant on wildlife matters.

Working closely with David during the anti-poaching campaign had enabled me to get to know him well, and I discovered that he was not the formidable character I had at first thought, but rather a strong, withdrawn and extremely shy, independent man, who took a lot of getting to know. His early upbringing had probably accounted to a large extent for this, for at the tender age of seven, he had been sent to boarding school in England, and circumstances, in the form of the Depression, had prevented him from seeing his parents again until the age of sixteen, when he left school and returned to Kenya. The tall youth who was reunited with his family after this long separation had, of necessity, learnt to be self reliant at a very early age, and had also learnt how to conceal his feelings and innermost thoughts behind a barrier of reserve, sometimes mistaken for arrogance. But, behind all this was a sensitive and compassionate nature driven by an almost fanatical dedication to his work and a rigid code of honesty.

Bill and I had now been married for five years, and we had to admit to the fact that during that time we had drifted apart. With dismay I had to confess to myself that I had made a mess of my life; that I should have heeded the words of my parents after all; that I had perhaps married too young. But, I also had to confess that my feelings for David went a lot deeper than mere friendship. I found in him a kindred spirit with interests and views so akin to my own, that I could almost predict what he was going to say. But any feelings he might have had for me were guarded and concealed behind an unapproachable wall of aloofness and reserve. He had also declared emphatically on many occasions that marriage was for the birds, and that he would never make the mistake of becoming entangled twice. He gave the impression that he regarded all women as pleasing playthings, but not something to get too involved with at any time. Even so, there were occasions when my woman's intuition whispered that I was perhaps in a different category and that he was not as totally indifferent to me as he appeared.

Another year passed, and still the gulf between myself and Bill widened. I found happiness and contentment in Tsavo, nevertheless, for I had grown to love the rugged country with its brick red earth and twisted, gnarled trees that had a special silvery beauty of their own even when stark and dormant in the height of the dry season. I loved the solitude and the space, the song of cicadas and the many bird calls, the sight of wild animals grazing peacefully below the house, and the contrast of the different seasons.

However, all this came to an abrupt end one day when a letter from Head Office to Bill offered him the chance of promotion and a transfer to the far away Mountain National Parks north of Nairobi as a full blown Warden.

Naturally, he was eager to accept the opportunity of advancement and excited at the prospect of a Park of his own, but to me it came as a bitter blow. The knowledge that I would have to say goodbye to Tsavo filled me with a numb misery. The crossroads now lay before me; which turning should I take? I realized that the decision that faced me would probably be the most momentous of my life.

Bill and I decided to go our own separate ways in the end, but it was a decision we didn't come to lightly. The future filled me with foreboding, but I decided to try and face it with courage and start again. So, while Bill proceeded to Nyeri, I remained in Nairobi, and took a shorthand typing job, while Jill started school.

Life in the city was very different to Tsavo, and I found myself mechanically pounding the typewriter with my mind and heart 300 miles away. However, an excuse to return to Tsavo periodically was conveniently provided by the presence there of my brother, who must have been rather surprised at how popular he had suddenly become, although he politely didn't say so at the time. I was glad to take advantage of long weekends and public holidays to escape the trammels of civilization and feel the peace of the bush again.

David and I corresponded spasmodically, although in retrospect, knowing what I do now about David's letter writing, it was frequently by his standards! We often met up, also, when he came up to Nairobi, but I accepted the fact that he was a confirmed bachelor, and I, myself, was not unhappy as I was.

During my absence, the development of the northern area continued, with the construction of a network of tracks, the completion of the Ithumba headquarters and the installation of several game blinds on the banks of the Tiva River, where one could sit in comparative safety and watch animals streaming down to the dry sand river to drink from the water in the holes excavated by elephants. Water is life in this arid land, and it was the elephant who made it accessible to other creatures; even the doves flocked in each morning to drink from the holes.

David loved the northern area because it was more remote and unspoilt. There is something very stirring about being alone in a wild place; something that seems to touch the very chords of our being, for man is a displaced animal, and subconsciously craves the natural life he has never known with a dull but insistent hunger that draws him back to such places. This need becomes more essential the more civilized we become, and people immersed in concrete have a definite yearning to escape from the shackles of civilization and lose themselves in time for a while. It is refreshing to sit quietly and alone in the bush, contemplating the stillness and the peace; listening to the humming of insects and the pure intonation of the birds, the whisper of the wind as it stirs and rustles the foliage in the trees and the low lapping sound of running water. It swells one's heart to look carefully about and acknowledge the marvels around us; the beauty of the rugged landscape, the delicate outline of a butterfly, the soft and ever changing shape of the clouds above in a sky of unbelievable blue; the fluid depth of a piece of quartz tinted with the exquisite colours of the rainbow, and the majesty and magnificence of the distant mountains. One can discover oneself anew in such a place, and things fall suddenly into the correct perspective. What before seemed insurmountable problems appear of little consequence beside all this. The solace derived from being in such places is certainly addicting, and having discovered it, one craves it forever more.

The game blinds on the Tiva were designed in order to try and capture this precious commodity for those that needed it. From the banks of the river, it was possible to witness a spectacle as old as time itself. When the moonlight played on the pale sand, dark shadows silently emerged from the night to take their places on the stage in front. The silence was broken by the sound of heavy breathing, ivory trapped the moonbeams and gleamed a silvery white as the elephants congregated. With infinite patience they began tunnelling down into the sand with their trunks until the water seeped slowly into the holes to enable them to drink.

As the night wore on, so the tempo increased, as rhino and other creatures took their places. Suddenly the sound of splashing water was accompanied by trumpeting, puffing, snorting, roaring and the pathetic pleading squeaks of the rhino as they strove to gain possession of a hole, and warded off intruders. Night's dark curtain was finally lifted, and as the sky became streaked with a fiery dawn, the players quietly withdrew to allow a brief interval of tranquillity, with only the fresh imprints in the sand and the pungent smell of the dung left to prove that it all was not a dream. Then the sandgrouse, doves and baboons appeared, followed by the shy kudu and other diurnal creatures, stepping cautiously in to slake their thirst from water exposeed by the elephants

On one occasion, David himself was able to be a part of this drama. He was standing in the riverbed contemplating the holes in the sand, when a kudu stepped down from a bush on the bank. It happened to be looking the other way, so David quickly crouched, pretending to drink from one of the holes in the same way as a baboon. Without hesitation or concern, the kudu walked up and drank deeply from another hole only a few paces from him. It was a very special sensation to find oneself accepted by one of the shyest of all bush dwellers. In that moment he was not an intruder, but simply another animal in need of a drink.

The southern portion of the Park, catering for the general public, had all the modern amenities in the form of signboards, ticket offices, notices, Lodges and Camping Grounds, which, although very necessary, tended to detract from the natural scene. One day, with the growth of tourism, long queues of minibuses would be entering the Park, flocking to the places of interest where hordes of people would be disgorged. Some people, already initiated into the thrill of wild places, would derive no pleasure from viewing game under such conditions; those who found wide highways and ticket offices, litter and the sight of notices, offensive in such places, and although they were a very small minority at the moment, David felt that their numbers would grow with time. With this in mind, he put forward a proposal that the northern area be afforded Wilderness status, and kept as wild and as unspoilt as possible, and that only safari parties under the supervision of a recognized professional guide should be permitted to camp there. By restricting the number of parties allowed in at any one time, and prohibiting day trippers, visitors who took advantage of this facility could enjoy and savour to the full the knowledge that they were well off the beaten track and had the area to themselves for the duration of their stay. For this privilege they would of course have to pay more than the normal tourist.

It was also suggested that a semi-permanent tented camp be established somewhere on the north bank of the river, to cater for those visitors whose means could not run to a professional safari. Such people would have to get themselves to a point on the south bank, and would then be ferried across the river. During their stay they would have to make use of the game viewing excursions organized from the tented camp, and in order not to conflict with any other campers in the northern area, their activities would be restricted to a certain portion of the Wilderness Area.

It was not easy to convince Head Office that this would be the best form of development for the northern area, but in the end they agreed. Ideally, the tented camp should have been established and managed by the National Parks themselves as another service to the public, thus avoiding the intrusion of private enterprise which always seemed to create problems and subject the Parks to pressures that were not in their best interests. However, Tsavo East had always been the 'Cinderella' amongst the Parks, and the authorities were not prepared to release the necessary finance for this enterprise. Finally a firm established by a well-known professional hunter, Glen Cottar, was granted the concession of establishing and running the tented camp, which became known as Tsavo Tsafaris.

CHAPTER 8

A New Life

NOT until my divorce was through did I discover that David cared for me sufficiently to overcome his cold feet regarding marriage. His proposal came out of the blue and took me quite by surprise, and after the initial shock I couldn't resist getting my own back in a small way by asking him what made him think that I wanted to take on another man having just got rid of one! His puzzled and worried expression turned to one of relief as I burst out laughing.

We were married quietly in Mombasa and round about the same time Bill, too, remarried; a happy conclusion for all concerned. Once again, I found myself coming to live in Tsavo East, but this time not with misgivings but with a great excitement, and I looked forward to the years ahead with confidence and eagerness.

Having settled in, once again I found myself David's typist and in addition to this I shouldered the responsibility for the accounts, relieving him of the one aspect of his job which he hated most. I accompanied him wherever he went, whether it was supervising the construction of a bridge, taking levels for a dam, shooting a maimed elephant, catching an orphaned baby, or merely carrying out a routine patrol.

Five thousand square miles is a very large area to have to cover, and of necessity a lot of our time was spent camping out in the remoter limits of the Park. David's years as a professional hunter had taught him how to make himself comfortable on safari and he insisted on a high standard. 'Any fool can be uncomfortable', he would say. Quite often, we were accompanied by VIPs, and although I too had camping experience, I found the responsibility of having to organize the provisions for such safaris weighing heavily on my shoulders at first. I had a dread of letting down the side by forgetting something vital, like the salt or the tea; the soap or the matches! However, having been guilty of this sin once or twice, I made a series of comprehensive lists that were pretty foolproof.

When the duration of the proposed trip was likely to be more than a week, the safari fridge, built into a stout box for travelling, came along as well, and this made housekeeping in the bush a good deal easier. In time, I became quite proficient at baking on safari, and for the purpose used an improvised oven made from an empty paraffin tin, fitted with a shelf half way and a lid. By heaping glowing coals around this and on top, regulating the temperature by adjusting the amount of coals, I was able to turn out fresh bread and even cakes, which always impressed our important guests. There is quite an art in cooking over the coals, and old Frederick, our cook, was an old hand at the game and very adept at it as well. Within minutes of arriving at our destination, a huge bonfire would be lit, and the glowing embers scraped aside, upon which the kettle of water was plonked. Invariably, just as the last tent had been erected, and when everyone needed it most, the tea would arrive with perfect timing.

Our camps were really quite civilized for we even had electric lights, the 12 volt bulbs being connected to the battery of the Land-Rover. One could have a hot bath, or a hot or cold shower depending on the weather. A brush enclosure served as bathroom, and the shower, a tin with a fitted shower nozzle, was filled and hoisted over a branch by means of a stout rope, while a bamboo mat placed underneath kept your feet clean.

We usually tried to site our camps by a river, and obtained clean water by digging a hole in the sand. The water that percolated into the hole had been well filtered first through the sand and was always a lot cleaner than that of the river itself. Occasionally we had to rely on water from a muddy water-hole, in which the elephants had been having a good time, and in this case the water had to be 'treated'. David knew of a root which when swirled in a drum of chocolate coloured water, had the same effect as a piece of alum. The mud slowly subsided to form a thick sediment on the bottom, while the water on top was comparatively clear.

I too had some old fashioned tricks of the trade up my sleeve, gleaned from my pioneering ancestors. Carrots and other root vegetables could be kept for quite long periods if 'planted' in damp earth, and a guinea fowl stuffed with charcoal and leaves and hung in an airy place without being washed in water, would keep fresh for several days.

Many people have a horror of camping in the bush, believing that every wild animal is bent on attacking them, but in fact, provided one takes a few elementary precautions such as refraining from putting one's camp in the middle of a game trail, or near a rhino's midden, and sleeping always under a mosquito net, there is really very little danger. I always placed great faith in the mosquito net, which was not only aimed at excluding mosquitoes and other creepy-crawlers, but snakes and lions too. Although the Tsavo lions had a bad reputation and probably constituted the most likely threat, we never even closed the flaps of our tent, for, to the best of my knowledge, no one has ever been dragged from under a mosquito net. The theory is that a lion will not attack unless it knows exactly what it is dealing with, and the nebulous appearance of the mosquito net provides an adequate deterrent.

The first night in a camp is the most likely time to expect trouble, when elephant and rhino might stumble on it unexpectedly, but once the presence of the camp in an area becomes known, and it does very quickly, then all animals are careful to avoid it. A good camp fire to advertise one's presence is usually the best protection.

Of course, the usual camp followers, like hyaenas and jackals, are regular nocturnal visitors, scavenging for titbits, but one must bear in mind that they are far more scared of you than you are of them, and would slink into the shadows immediately you appear out of the tent. The same thing goes for snakes, which, although extremely numerous throughout the Park, pose very little threat, and will always attempt to get well out of the way if they possibly can. Puff-adders, being more lethargic than other species, are probably the most dangerous of the poisonous snakes in that one is more likely to tread on them inadvertently. Again, a few elementary precautions taken when a camp is established, or around the vicinity of a house, minimize the risks. One has merely to keep the grass short and clear away brush and dead logs which might afford cover. The chances are that the presence of a snake will be detected by the birds, who will always give warning by congregating nearby and twittering excitedly, and sometimes even swooping onto the snake repeatedly. When one lives in the bush for any length of time, one automatically learns to interpret the language of nature; the barking of a bushbuck might denote the presence of a leopard or a lion; the chattering of tickbirds means that a rhino or a buffalo is lurking nearby; the sight of vultures sitting in a tree rather than on a kill indicates that the lion is not far off, and in this way, if one keeps one's ears and eyes open, one can usually succeed in circumventing trouble. In addition to this, when one knows animals really well, one can predict their reactions with a measure of accuracy. One can tell from the way it behaves whether an elephant really means business, or is merely bluffing, and at first I thought David used to take outrageous risks and would reprimand him for doing so. He would simply laugh at my fears and explain why, in fact, there had been no risk involved. For instance, a cow elephant with a very small calf, is unlikely to charge and leave the baby exposed to danger; a young 'teenager' bull might put on a convincing show of aggression, but lacks the confidence to do much more.

There are instances when a lightning decision must be made in order to extract oneself from a tight corner; occasions when only a moment's hesitation could mean losing the initiative and end in disaster. One such occasion occurred when we topped the bank of a tortuous *donga* to find an old mean-looking cow elephant waiting for us in the middle of the track at the top. There was no turning back and there was also no mistaking her evil intentions either, so David slammed his foot hard down on the accelerator, and literally charged straight at the elephant, not hesitating even for a split second. Horrified, I held my breath, while the elephant dropped her enormous head to take the impact she thought was coming, backing away a few paces in the process, her ears out like two gigantic sails. Those few paces backwards enabled us to shoot past by the skin of our teeth, swerving violently at the very last moment, and almost brushing the cow's head as we did so.

When the elephant realized what had happened, she screamed and trumpeted her rage, and proceeded to pound down the road after us, but having passed her we now had no difficulty in out-pacing her.

The excitement of this encounter left me looking like a ghost, and quite speechless for several moments. My heart was thundering against my ribs so violently that I thought I would pass out at any moment. It was the closest shave I had ever had!

In fact, the story didn't quite end there, although at the time we thought it had. Several miles further along we happened to get stuck in a sandy crossing, and were leisurely gathering logs and brush to place beneath the wheels, when, to my horror, our old friend turned up again on the far bank, still in full cry! This time, we were sitting ducks, and I was rooted to the spot almost paralysed with fear, while David hurled a pebble at her and clapped his hands. After a few short rushes which terrified me even further, she thought better of it and ambled off, to my immense relief, needless to say.

I must own up to being somewhat scared of elephants, for although they are tolerant and peace-loving animals, they are possessed of incredible strength and can be terrifying at close quarters. I like to blame David as being responsible for my fear of elephants and accuse him of having broken my nerve by scaring the wits out of me so often.

The first five years of my marriage are threaded with recollections of those first orphans we acquired described in *The Orphans of Tsavo*. A kaleidoscope of memories of many creatures of differing species, who shared our lives and our home, and who all contributed to our knowledge. No one can really know an animal until they have actually raised a baby of that species, for enjoying an animal's implicit trust and taking the place of its mother, one gets an insight into many hidden characteristics of its kind, and will come to look upon its wild brethren in a new light. It is possible to acquire from books a certain amount of information about the habits of a certain species, but this has little impact, being impersonal and dead. When one has kept a member of that species, and shared in its development, its playful moments and its moods; been the object of its affection and its trust, and can interpret its every sound and expression, then one begins to understand something about that particular type of animal. One will learn also, should there be another, that individuals of the same species vary considerably in character in exactly the same way as do humans. Having accepted this, one will have learnt one of the most basic lessons which many humans, due to their arrogance, are inclined to overlook; and that is that every animal is an individual in its own right and should not only be recognized as such, but have allowances made for a certain amount of individual expression and choice in the matter of its habits.

But of those early orphans only Samson and the two rhino, Rufus and Ruedi continue into this story to pick up the orphan theme.

When Samson was rising twelve years old, he was quite an impressive animal, well over eight feet tall, with a massive frame and fair-sized tusks. He had developed a strong attachment to Rufus, that gentle, sloppy rhino orphan that had been with us from the day of his birth. They used to play together frequently; a most amusing sight, for, in order to be on an even footing for the game, Samson used to either kneel down, or lie on the ground, warding off the onslaughts of Rufus with his trunk and very little effort.

All our past female orphaned elephants had no problem rejoining the wild herds, for elephants are very sociable towards one another and will always welcome a stranger into their midst if it is another female, or a young bull. Bulls of Samson's age, however, find themselves rejected by the cow herds, and even in the wild state have usually by that time left their family unit.

Elephants practise a matriarchal system. The population is comprised of family units under the leadership of the dominant cow of the family, with her daughters and their offspring forming the unit. This little group is an independent entity, although they may at times loosely attach themselves to other family units to form larger herds, usually when conditions are good and food plentiful. Loyalties in the family units are binding and lasting, and under normal circumstances members of the unit may spend their entire life together, until the leader, by then possibly a great-grandmother, dies of old age round about sixty, when the last of her six sets of molar teeth has been worn down. Her role as leader is then filled by the dominant daughter, and so on.

Bulls have no permanent place in the cow herds, the bull calves being driven from the unit at the time of puberty, to roam individually or in small bachelor groups, severing their family ties completely at this stage. Adult bulls lead an independent existence, joining the cow herds on a temporary basis only, owing no particular allegiance to any one unit or any one cow, going their own way again at will. There is evidence also that the bulls are inclined to be more attached to specific ranges than the female herds, keeping to areas that can be possibly classified as 'bull areas'.

Due to the social structure of elephant society therefore, life was not so straightforward for Samson, and by the time he felt that he would like to rejoin his own kind, he had left it too late to qualify for entry, and he learnt some lessons in elephant etiquette the hard way.

There came a time when he would become very excited when scenting wild herds, no doubt aware of the proximity of a lady in an interesting condition, and he would eagerly rush off to investigate. On occasions he would be away for a month and more, and just as we were beginning to think that he had made the break at last, he would reappear; footsore, dejected and obviously greatly disillusioned, having been put in his place by the wild elephants for being so presumptuous. Very often he carried evidence of this in the form of tusk wounds in his flank and rear end.

Although we had no desire to be rid of Samson, who, after all had been part of the family for twelve years, we knew for his own good that the time had come when a break had to be made, for he was becoming a problem around the headquarters, turning on taps and not turning them off again, breaking down trees and gobbling up people's gardens. And so we decided that we would have to forcibly evict him, in order to encourage him to sever his human ties and return again to where he rightly belonged. It was very difficult for David to harden his heart and drive Samson away from the place where he had always found comfort and shelter, and it was doubly difficult to do this in view of the close friendship he and David had always shared. Ever since the day when Samson had been captured, and David had subdued him and won his trust, there had existed between them a special affinity. It was not easy to break that trust. However, it had to be done, and finally the time came when the break was complete, and Samson avoided the headquarters, roaming the Park like his wild brethren, sometimes alone, sometimes in the company of other bulls, until he happened on the shady reaches of the Galana River near Lugard's Falls, and there he decided to stay. To this very day Samson can often be seen there, peacefully feeding on the lush vegetation bordering the river.

We have often longed to go up to him and renew that long friendship, but at the same time we realize that such an approach would not be in his interests, for although Samson will never have the fear of humans that his wild brethren acquire at a very early age, it is essential, for his own safety, that he learns they are not to be trusted. Man, detested by all wild members of the animal kingdom, although once this particular elephant's closest friend, happens also to pose his greatest threat, and it is because of this that we have not encouraged the friendship for our kind which Samson would willingly give.

With Samson's departure, the tradition of Tsavo East's elephant orphans was carried on by 'Eleanor', a particularly endearing female elephant, who was actually not a native of Tsavo, but who had been found by Bill abandoned in the Northern Frontier Province where he had been conducting a safari for the Governor of Kenya and his wife. Eleanor was named after Lady Renison, the Governor's wife. She had been kept for a time at Nyeri and then moved to the Nairobi Park Animal Orphanage, but neither place had suited her, and she had suffered a lot of ill-health as the result. Because of this, it was decided that she be moved to Tsavo East.

Eleanor was about five years old when she joined us, and even at that early age showed signs of developing into a particularly gentle elephant, possessed of a very strong maternal instinct, for she liked nothing better than being able to adopt a newcomer, fussing around it, protecting it and mothering it with such tenderness that it might have been her own calf.

She stepped into the role recently vacated by Samson as leader of the Orphan Herd, and her 'unit' was comprised of one other younger elephant, Kadenge, brought home by Samson, who had come across it in the Voi Forest; Rufus and another rhino calf a year or so younger called Ruedi, several buffalo and three ostriches. Each evening, after a day spent browsing in the Voi River valley, this rather unusual assortment of different animals would file back to their stables, headed by Eleanor, who took her responsibilities very seriously, and insisted on her rightful place at the head of the column.

It had taken quite a long time to persuade Ruedi to come to terms with Rufus, for true to rhino form, they had hated each other at first sight. However, with perseverance and after many arguments, they had become firm friends in the end, but, like all friends, there were times when they had a difference of opinion. These disagreements usually started in fun, but when one poked the other a little harder than the rules allowed, it ceased to be a game and ended in quite a serious confrontation. Eleanor, being peace loving and gentle herself, detested these squabbles, and always intervened to break them up as soon as possible. She would trumpet and charge in between the two contestants, flailing them with her trunk and poking at them with her tusks until they broke off the engagement, and she would then separate them well and truly by chasing them in different directions. This unruly behaviour on the part of the rhino was probably responsible for her acquiring rather a dislike for rhino generally, which became more evident as she grew up.

Tsavo suited Eleanor. She thrived and overcame the ill-health that had plagued her upcountry, and she accepted her responsibilities with a maturity few human children of her age could begin to match.

Kadenge had never been handled or tamed – he had merely arrived one evening having followed Samson back to the stockades, where he plonked himself for the night as though there was nothing unusual in this. Samson had been rather touched, I think, for he was very attentive for the first few days, but soon lost interest and simply accepted Kadenge's presence in the same indifferent way that he treated the other orphans.

Kadenge had a definite mischievous streak in him which became very apparent whenever a minibus full of tourists drew up to photograph the orphans. He would nonchalantly wander off a little way and pretend to be feeding until the time came when all the passengers were about to clamber into the bus. Invariably, at this moment, he would give vent to a shrill trumpet and bear down on the bus like a ship in full sail. The bus driver, thinking that he was being charged, would roar off, wheels spinning, in a cloud of dust, some tourists hanging on by the skin of their teeth having only just managed to get in in time.

There was no doubt at all that Kadenge derived great enjoyment from putting the bus to flight, and after pursuing it for a short distance, would take possession of the road, swinging his head round this way and that with ears widespread in an arrogant and aggressive way.

This rather unfortunate habit became Kadenge's favourite pastime, and was repeated daily whenever some tourists arrived. It wouldn't have been so bad had he confined the chase to minibuses, but he took to doing the same thing to pedestrians, which also provided good sport. He even tried it on me once as I happened to be walking down to the office. The next thing I knew, Kadenge came hurtling down the hill, a most formidable sight to say the least! It took all my willpower to prevent myself from taking to my heels, but I managed to stand my ground, and Kadenge, completely nonplussed, applied all four brakes at the very last moment, and came to an abrupt halt in a cloud of dust a few paces from me. Looking rather foolish, he turned round and retraced his steps, glancing back over his shoulder now and then as though to say, 'Spoil sport!'

The labour soon began to complain, for they were usually selected as the target, and so it was decided that Kadenge must follow in Samson's footsteps and confine his future activities to chasing his own kind! The task of driving him out was comparatively painless, for we had not made an attempt to befriend him particularly. Because of this, all that was required was one or two thunderflashes to make him understand that he was 'persona non grata' around the headquarters.

No doubt, somewhere in Tsavo, a minibus is charged by a young bull elephant for no apparent reason, and Kadenge, living free, can still derive some pleasure from his favourite sport and make the visit a memorable one for the tourists!

Eleanor was not without a 'baby' for long, fortunately. One day she happened to find a lone youngster in the bush which she proudly brought home that evening, just as Samson had done in the case of Kadenge. We suspected, judging from the barrage of shots we had heard the previous evening, coming from the direction of the Sisal Estate, that this little elephant's mother had been shot on control, having been guilty of damaging the sisal.

CHAPTER 9

The Changing Face of Tsavo

WHEN the Park was first set aside, the vegetation of the eastern section had consisted almost totally of very thick *Commiphora* woodland. Below the tree canopy grew the tall *Sansevieria*, a wild sisal which formed a formidable barrier through which thin game trails crept like a spider's web, linking the water-holes to the feeding grounds. 'Woodland' was an apt description for the vegetation in those days, for the twisted commiphora with their characteristic flaking, paperlike, blue and green bark, stretched interminably in all directions to form dense cover. Scattered amongst the commiphora were other types of trees; tall evergreen *Melia volkensii*, the statuesque *Delonix elata*, giant baobabs, flat, umbrella shaped *Acacia tortilis*, and many other varieties. Riverine types fringed the rivers and luggas in the Park; tamarinds, doum palms and *Acacia elatior* amongst others.

Although the Tsavo Park had not been created due to a preponderance of wildlife, but rather because it was the only area that could be relinquished without conflicting with human interests; by fortunate accident, the Park as a whole happened to incorporate two distinct faunal types; the Somali or northern races of fauna occurring in the arid habitat of Tsavo East, and the Masai or southern species represented in the lusher western side. More by good fortune, therefore, than by good judgement, the Park harbours a greater variety of large mammals than any other Park in the entire world. But, apart from elephant and rhino, none were found in any significant numbers. Kudu, dikdik and gerenuk were perhaps fairly numerous, but the sight of a herd of buffalo was a conversation point, zebra were only occasionally glimpsed in small scattered groups of never more than a dozen animals, impala found only on the Galana and Athi Rivers and Peters gazelle and the Somali ostrich confined to the northern area and never seen south of the Galana River. Waterbuck frequented the watercourses; eland were only very occasionally seen, again in small herds, and the rangy unpredictable Tsavo lion could not afford the luxury of large prides, but struggled for survival singly or in pairs. Oryx were perhaps the most numerous of the antelope and could occasionally be seen in fair sized herds.

By and large, those early visitors to the Park couldn't expect to see much more than elephant or rhino, and the odd smaller creature only if it happened to be standing on the road itself, so dense was the bush, and the Tsavo Park couldn't begin to compare with other places like the Serengeti in Tanganyika, or the Mara country and Amboseli. Although we defended it fiercely against its many critics, it was obvious that unless the number of tourists in years to come justified its existence, the long term survival of the Park was in jeopardy. Something had to be done to try and increase the stocks of game.

An attempt was made to open up some areas by burning certain stretches near the roads. In this way, it was hoped to try and encourage smaller animals to congregate and give the grass a chance to thicken up, but it was an abortive exercise, for the bush simply would not burn. It seemed as though one had to accept that Tsavo East would always be essentially a 'big game' Park, and although it contained a great variety of animals, none, apart from elephant and rhino, would ever be numerous in this type of habitat.

Round about 1956, however, a change started to take place, and the elephant took to felling the commiphora trees over huge areas of the Park. It is the nature of elephant to fell trees, and although commiphora are particularly vulnerable being easy to push over, the destruction that was taking place was alarming.

To begin with, this departure was attributed to the rather dryer than normal conditions that had been prevailing over the past couple of years, and it was thought that the trend would cease given a good year. But this was not the case. Indeed, it almost seemed as though the wholesale extermination of commiphora trees was a policy decision taken by the elephant generally, for it was clear that not only a quest for food motivated this practice. Usually the elephants ate very little of the tree once it was down, and very often never even touched it at all. Even in the wet season when there was a preponderance of food for the elephants, the commiphora trees were pushed over for no apparent reason.

By 1959 many areas of the Park were beginning to resemble a battlefield, or a lunar landscape, and the destruction was worrying David. He recorded it in many of his reports: 'During the past few years, the destruction of vegetation by elephant has reached serious proportions. If present trends continue, it is doubtful if the Park can continue to support the existing population much longer. What effect this will have on other species remains to be seen, but I think it is important that we should seek scientific advice regarding this problem as soon as possible . . . The destruction of vegetation in the Tiva valley is very noticeable, where hundreds of baobabs have been felled by elephant in their search for food.'

Thus started the 'elephant problem', which was to prove the source of much controversy in the years ahead.

It was round about this period that David made the acquaintance of Dr Frank Fraser Darling. An eminent ecologist, reserved, wise and cautious, Dr Fraser Darling possessed that rare and intangible quality that eludes most scientists. He had a feeling for nature, and a great love for it as well, making him not only an ecologist but also a very able naturalist. One could almost call it an instinct that comes with many, many years of experience and a close affinity for wild places.

It was Dr Fraser Darling who was responsible for opening David's eyes, for he demonstrated the value of careful observation of even the most commonplace things in the environment, which, when interpreted by a code of simple common sense, and pieced together step by step, provide the clues which lead to a proper understanding of a problem, and the emergence of a picture, often so obvious and plain that one is amazed that it escaped notice before. The surest road to the right answer usually lies along a simple path uncovered by commonsense, and because of this, the most obvious clues can easily be overlooked.

However, the ultimate result of the destruction of the commiphora trees was not clear, for it was feared that the rainfall of Tsavo East was inadequate to promote and sustain perennial grass cover, and such large scale destruction of the dominant vegetation type under these circumstances might prove disastrous. If not grass – then what? The general consensus of opinion was that it might be the emergence of a desert.

The position did look critical. Skeletons of trees lay in tangled heaps on bare baked soil, and even David's faith in nature began to falter. The one bright spot was the phenomenal powers of recovery of the dry country, of which few people had experience, for once rain fell, it seemed that the plants were specially adapted to make the most of the short growing period allotted them in areas of marginal rainfall. Attention was drawn to this in another report to Head Office: 'Elephant continued to create havoc with the vegetation in the Tiva valley, and there was hardly a tree or bush left undamaged. Heavy rain, however, altered the situation almost overnight and the bush has recovered in a remarkable manner. Shrubs that have been broken off a few inches from the ground sent up new shoots which reached a height of 2–3 feet in a few weeks, and the larger trees which appeared to be dead have nearly all sprouted again and produced vigorous growth.'

But, despite this, the bush could not compete and huge areas bordering the Galana, Athi and Tiva Rivers continued to be devastated by the elephants. By now, we were extremely concerned, and the advice of several scientists had been sought. All agreed that the elephant population should be drastically reduced in numbers before they annihilated themselves and other species as well.

It seemed ironic that it should come to this, when so much effort had gone into the anti-poaching campaign to stop the slaughter of elephant. The idea of shooting large numbers of these animals was abhorrent in the extreme, and contrary to all the principles of a National Park, which should, above all, be a sanctuary. It would also create a very dangerous precedent, but if the alternative was a desert, then there was no other course, and it had to be done. The words of Dr Fraser Darling came to mind. 'When it is necessary to kill an animal for science, food, or self-defence, be it harmful insect, beautiful bird, or mammal, think well first, but when the act is before you, emotion should be laid aside. Do not hate, love, gloat nor regret; kill well and cleanly and be done with it.'

What sound advice this was. And when David recommended to the Trustees of the Kenya National Parks that elephant be cropped in Tsavo East as a matter of urgency, more of Dr Fraser Darling's sound advice was foremost in his mind. 'Those of us who are prepared to accept the necessity for killing will not fail in our respect for life by being prepared to do the killing ourselves.'

But it was not only for this reason that David emphasized the importance that the cropping be undertaken by National Parks personnel. He was very conscious of the danger of allowing private enterprise to have any influence over a situation which should not be subjected to economic pressures and profit motives, but governed only by what was in the best interest of the Park itself, and what was right.

In those days, not a great deal was known about elephant;

their food preferences, social behaviour, herd structure, and movements were all aspects which had an important bearing on the problem. David spent a lot of his spare time trying to find the answers to some of these questions, and in this our elephant orphans played a valuable role. Many hours were spent with them, collecting and pressing samples of the plants they favoured which were then forwarded to Nairobi for chemical analysis and positive identification. This important groundwork would provide a baseline for the scientific team he hoped would one day form part of the Parks organization.

Africa is a land of extremes where droughts and floods are as commonplace as the summer and the winter in temperate climes. Although the average rainfall over most of Tsavo East is between ten and fifteen inches per annum, the rainfall probability maps clearly showed that approximately once in every ten years, a dry tongue that envelopes the arid Northern Province of Kenya extends south east and encompasses practically the entire Park. At such times less than ten inches of rain per annum can be anticipated, so the drought that gripped the Park in 1960–61 was not entirely unexpected, but it was particularly severe in the area west of Lugard's Falls, where the desolation along the river was terrifying. To make matters worse, large numbers of rhino began to die in this particular area.

We established a camp in the affected area, so that we could be on the spot to do all we could to alleviate the suffering of the stricken rhino. By the use of mobile pumps and sprinklers, water was pumped from the Galana to irrigate stretches along the banks in an attempt to save as many rhino as possible by promoting a green flush of vegetation, but it was like a drop in the ocean, and every day weak and dying rhino were found, who were beyond any help. The kindest thing was to hasten their end with a merciful bullet.

Because there was so little in the way of food for the rhino along the river itself, which looked more akin to a lunar landscape at this time, it was decided to try and pump water to storage tanks constructed on the top of the Yatta Plateau, and pipe it down the other side and some twelve miles inland by gravity. It was hoped that the provision of water in this area would relieve the congestion on the river itself by making good grazing available, which at present could not be utilized in the dry season. Teams worked feverishly day and night to install the headworks for this ambitious scheme. A trench for the piping dug up near the vertical and extremely rocky face of the Yatta was no mean task either, and progress on this project was not exactly helped by the arrival of hordes of journalists and photographers, who flocked to the scene to get sensational pictures. They often turned up unannounced, having persuaded Head Office in Nairobi that the cause of the rhino would benefit substantially from the publicity given to their plight. More often than not, although they were supposedly self-contained, they ended up by having to share our camp and our dwindling provisions as well, and all the time, David, who is not very partial to publicity at the best of times, and who already had quite enough on his plate, quietly seethed.

There was one party who happened to be travelling behind us when we came across a dying rhino lying exhausted in the blazing sun, too weak to get to its feet even when approached. As it struggled feebly, emitting pathetic cries, one jubilant reporter dashed up alongside David, who was standing looking down on the wretched animal, his face a grim mask concealing the pity that surged within him.

'Say!' said the reporter, excitedly. 'This is just great! Mind standing alongside the beast for a shot?'

David was at the end of his tether and glared at him in disgust. Without a word he raised his rifle and with one quick bullet in the brain, ended the animal's suffering, and the reporter's chances of a sensational picture! He then strode back to the car, slammed the door, and drove off, leaving the newsman gaping in astonishment and stammering protests about the lack of co-operation he had received.

Yet another irritation was a case of poaching in the Park, and the culprits this time turned out to be a faction of the Field Force itself. One of the Gate Rangers had reported having discovered the burnt carcase of a rhino near Sobo, and this, in itself, merited suspicion. On investigation, pieces of bullet were found amongst the charred remains, and it was pretty obvious that the Section who had been recently patrolling that area were in all probability responsible for the death of this rhino. By this time, they had moved to another area, and so no immediate action was taken until they were due to be re-called to headquarters, when the lorry was intercepted on the road and directed immediately to the headquarters. The patrol leader, who had been told to report personally, saluted smartly.

'Have you anything to report?' enquired David.

'No, sir.'

'You haven't come across any signs of poaching or recovered any trophies?'

'None at all, sir,' was the reply.

Their kit was then searched on the back of the lorry, and there, hidden in a cardboard box beneath all the tents, were two bloodstained rhino horns.

The three Rangers suspected of being implicated were arrested and placed on remand pending further police investi-gations into the case, but events took an unexpected and tragic turn. Another Ranger in the Section, who happened to be from a different tribe, volunteered certain information regard-ing the incident, and then promptly committed suicide by hanging himself in the lines that night.

It was indeed regrettable that the record of the Field Force, which until now had been without blemish, should have been tarnished by this extremely serious offence. No mercy was wasted on those responsible. They were convicted, jailed and discharged from the service forthwith.

Meanwhile, the death toll of the rhino in the drought-stricken belt along the river had mounted to over 300; an alarm-ing figure to say the least. One wondered if there would be any survivors at this rate, and the necessity for action had become more urgent. Many experts were called in to give advice on what should be done regarding the elephant. At this time, however, it was not known exactly what figure the elephant population of Tsavo East stood at, for no aerial census had ever been attempted. David hazarded a conservative guess at approximately 5,000, and after prolonged discussions it was suggested that about one third of the population should be removed within a period of two years, which would mean destroying just under 2,000. Everyone agreed that the data available were insufficient on which to base a cropping scheme, and that more statistics should be sought before a final decision could be taken. The most urgent need was a comprehensive count of the elephant population of the Park.

The help of the Army was enlisted for this task, and three Beaver aircraft plus two helicopters were placed at our disposal. The Park and adjacent areas were divided into blocks and maps made of the individual blocks, upon which observers plotted the number of elephant counted in each particular block. It took a week to cover the entire area, and advantage was also taken to obtain as many aerial photographs of the elephant herds as possible, so that some idea of the composition of the herds would be known.

All the data from the count, plus the photographs, were handed over to a well known statistician, Mr James Glover, whose interest in wildlife and elephants in particular prompted him to give much of his spare time to a study of the problem. He devised an ingenious means of reducing the elephant population by one third in a comparatively short space of time, by the selective cropping of a small number of animals from the younger age groups, which formed the reservoir for future recruitment. But the practicability of this suggestion posed difficulties that were insurmountable; the problem of finding people capable of ageing elephant calves in the field sufficiently accurately to select the correct age groups; the disturbance such a measure would evoke amongst the herds themselves, in all probability turning the adults savage and a danger to the visiting public, or driving large numbers out of the Park into neighbouring African settlements, with equally disastrous results. There seemed no solution to these problems, but when the heavens opened towards the end of the year resulting in countrywide floods and such profuse regrowth that the former lunar landscape faded into a tangled Garden of Eden, the elephant problem fell into abeyance once again and seemed to lose its urgency.

The lazy river overnight turned into an angry, boiling, brown torrent that poured hungrily over its banks, uprooting thousands of beautiful *Acacia elatior* and poplar trees, and inundating the many reed covered islands in midstream. Work on the headworks to the pumping scheme had to be hastily abandoned, as did our camp by the river, and the following morning the place where the tents had stood was completely covered by the floodwaters.

It was a joy to see the desolate landscape turn to a lush and vivid green, the once bare earth become carpeted with wild flowers of every hue, creepers festoon the fallen trees and low shrubs in a delicate lacework of leaves and suddenly burst into a shroud of papery, pure white blossoms, making the countryside appear as though covered in snow.

What a strange part of the world Tsavo was; a land of such dramatic contrasts and differing moods; one moment a barren desert, then a painted paradise, now a still, dormant, hostile wilderness, then suddenly vibrant, and alive, abuzz with the activity of millions of insects, and ablaze with the colour of an unbelievable variety of wild flowers, around which danced hundreds of fragile little butterflies, each one a masterpiece of Nature. It was the extremes in Tsavo which gave it so much magic. Even the rhino, which yesterday had been more dead than alive, were soon cavorting happily in mud wallows, while the elephant romped with carefree abandon in the rainfilled pools, mirroring the mood of the bush.

All the waterholes were soon brimful, while Aruba dam roared over both spillways to form a shallow lake below in the river valley, a paradise for water birds; wild duck, geese, storks, cormorants and ibis, as well as frogs of all shapes, sizes and colours, whose differing voices set up a deafening symphony that could be heard for miles around. The pure white frothy substance containing the eggs of tree frogs adorned each bush overhanging the waterholes like baubles on a Christmas tree and the air was almost heady with the scent of myriads of mixed blossoms. It seemed as though all creatures, whose sole preoccupation had been mere survival, now cast off this heavy burden and could afford the luxury of love and of play.

The creeper, *Ipomoea mombassana*, which covered practically the entire Park during this period, and particularly the areas where the vegetation had suffered most punishment during the drought, was, in fact, playing a very important role although, of course, this was not apparent at the time. This creeper usually appears briefly after each rainy season, but never before had anyone seen it so profuse, and where all the millions of seeds came from to produce such regeneration, was a mystery. One can but suppose that in years gone by, a similar situation had arisen, and the seeds had lain dormant in the soil, awaiting a recurrence of optimum conditions in which to germinate.

Not only did the *Ipomoea* provide an immediate and ready source of food for most creatures, and particularly the elephant, considerably easing the pressure on the trees, but it acted also as a bandage, covering the beaten countryside, providing shade and cooling the soil below. Given these conditions, slowly but surely, grasses began to emerge; not just the annuals one expects during the rains, but valuable perennial grasses, while the presence of the fallen commiphora trees, acted as an effective barrier against trampling by heavy animals, and although dead, continued to fulfil a function and contribute to the healing process. By being afforded this protection during the early critical stages, the grass seedlings were able to become established, seed and spread out, boosted by an additional supply of water in the form of the run-off from the twisted trash above. Meanwhile, the termites were busy enacting their role, steadily removing the dead wood, carrying it back into the ground for their fungus gardens, not only disposing of the spoil, but enriching the soil in the process.

By far the most serious feature of the floods appeared to be the destruction of most of the riverine trees, for, in many places, the river had widened its course by a hundred feet and more, taking with it large stands of beautiful *Acacia elatior*, doum palms, poplars, tamarinds and wild figs, and leaving long stretches completely devoid of vegetation. But Nature compensated even for this catastrophe, turning it into a blessing in disguise. For some years, the experts had been troubled by the lack of any sign of regeneration of the *Acacia elatior* trees along the river, many of which had been ringbarked and killed by elephant during the drought. It was feared that this particular species might disappear completely from the Park unless afforded special protection, but the floods must have provided just the right conditions, covering the banks in a thin layer of silt, for suddenly *Acacia elatior* seedlings popped up in their thousands. In years to come these trees would fringe the river in bigger and better stands than before and would not disappear forever. The same was the case along many of the dry watercourses, which had also been subjected to floods, and in some places veritable forests of young acacias appeared. We were amazed by this sudden turn of events, until David remarked that the adult trees seemed to occur in distinct age groups anyway, and so obviously this was not such a phenomenon after all, and had very likely happened hundreds of times before.

Several years of average rainfall followed the 1960–61 drought, and during this period large areas of former commiphora bushcountry became colonized by the perennial grasses, whose value in terms of productivity was far in excess of the former régime. Incorporated in this grassland was a wealth of valuable legumes favoured by many of the browsing animals, including rhino and elephant, and one couldn't help but wonder whether perhaps the conversion of some areas of Tsavo East to an open savannah type habitat was, in fact, such a disaster? By now it was obvious that the desert, people feared would replace the commiphora woodland, was not part of Nature's design. They had doubted that the rainfall was adequate to promote perennial grass cover, yet it had come about. Some argued that it couldn't last, but only time would answer this.

Many experts on grasses visited the Park, and all expressed surprise at what they found. In fact, one man, a leading authority on the subject, could not bring himself to believe that the area had so recently been woodland, until he was shown the remnants of sansevieria amongst the grass, and had then to concede that this could only have become established in a thicket environment. The Agricultural Department even sought permission to collect the seeds of certain grasses in the Park, so that these could be despatched and sown elsewhere. It seemed as though Nature had confounded the experts.

The tree skeletons, of course, were still very conspicuous and tended automatically to draw the attention of a layman, often leading to adverse criticism and comments about the appalling destruction. But those who looked a little closer and could see beyond the débris, realized that the term 'destruction' could not really be applied, for scattered in amongst the grass cover over wide areas was the dark green foliage of young *Acacia tortilis* seedlings taking the place of the commiphora trees. A transition was obviously taking place, and it was true to say that the vegetation was not being 'destroyed' but 'changed'.

Quite obviously, this development would have far reaching consequences on the Park, for changes in the flora would most certainly affect the fauna as well. Where some species would gain, others would undoubtedly lose; such see-saw cycles were a part of Nature. But by and large, the indications were that the changes would prove beneficial to the majority of species. There was likely to be a rapid build-up in grazing animals with the appearance of a more open habitat, but the changes might prove detrimental to bush-dwelling creatures, particularly the lesser kudu, although there was no reason to suppose that this species was in any danger in the foreseeable future, for large tracts of the northern area remained untouched. The bush along the luggas and watercourses would also continue to afford cover for these shy creatures, and from the diet angle, they were known to be extremely adaptable.

Grassland could support a wide range of herbivores, which would in turn result in an increase in the numbers of carnivores,

and the game would be a lot easier to view in more open surroundings; an important consideration with the paying public.
Another comforting thought was that whereas the best game
areas in East Africa were under an acacia grassland régime,
the same could not be said of commiphora woodland country,
and nowhere did this type of habitat support a sizeable game
population.

It was at this stage that Tsavo East acquired an aeroplane,
a Piper Supercub donated jointly by the New York Zoological
Society and the Kenya Government. David qualified for his
Private Pilot's Licence, and brought the plane back to Voi. It
proved invaluable in a hundred ways. Not only was it now
possible to do before breakfast what before would have necessitated a two day safari to the northern limits of the Park, but
it enabled him to watch the vegetational changes very closely
and plot the movements of elephant. It was also used for anti-
poaching work to direct the ground patrols from the air by
radio, for locating ivory lying in the bush and leading patrols
to the spot so that the tusks could be recovered, for selecting
the best alignment for a new road, co-ordinating men on the
ground in the event of a bush fire, or simply getting from A to
B in a hurry.

The conversion of the habitat in itself was strange enough,
but it was accompanied by an even more amazing development; the gradual appearance of over a hundred permanent
springs and seepages within the Park. Water bubbled from
many unexpected places along the Yatta escarpment, while
formerly dry watercourses seemed to suddenly stir from their
long inactive sleep to trickle as little streams which increased
in volume until some boasted pools large enough to accommodate hippopotamus. Even unlikely looking dry dongas in extremely arid settings many miles from any other source of
water suddenly displayed a suggestion of dampness, and with
the activity of elephants turned into a small pool, and in
time began to flow, first only a very short distance, then a little
further, until eventually a series of fairly large intermittent
pools linked by running water heralded the birth of yet another

stream bringing life to this thirsty land. We could hardly believe our eyes! Every time David flew over the Park a new spring seemed to have appeared, but he didn't dare think that it could last, and assumed that the heavier than usual rains that broke the 1960 drought must have been responsible for a temporary rise in the water-table. There was however one aspect of all this of which he was convinced; that these same watercourses must have flowed before as streams in the distant past, for it was extremely unlikely that the huge *Acacia elatior* trees and doum palms which fringed the banks, could have been able to become established under any other circumstances, being essentially riverine species.

Meanwhile, research into the elephants of Murchison Falls and Queen Elizabeth National Parks in Uganda was being undertaken by a team of scientists headed by Dr Richard Laws, who had decided that their over-population problem warranted the cropping of some 2,000 elephants. Ian Parker's newly formed company, Wildlife Services, acted as executioner. The cropping was extended to hippo as well, which was also undertaken by Ian and his team. And although we were glad to hear that Ian's new enterprise was flourishing, and appeared to be going from strength to strength, the belief at the back of David's mind now turned to a firm conviction; that if cropping in a National Park was indeed the best course to adopt under certain circumstances, it should not be undertaken by private enterprise, particularly when the profits were so substantial.

As far as Tsavo East was concerned, the nagging doubts about the best course to adopt persisted in the light of trends since the floods; the perennial grasses that had colonized the former bushcountry so rapidly, the evidence of a marked increase in plains animals, the more open and pleasanter surroundings and easier terrain, the incredible recovery of the country that had suffered such punishment during the drought, which not only now looked in good heart but continued to support an extremely large, healthy rhino population despite the heavy mortality of 1960; even the mysterious springs and streams; all these factors fanned the doubts in David's mind about the wisdom of tampering with the elephant population at this stage when there were so many facets of the problem that were not fully understood. It could be very damaging to the Park to take the wrong road at this critical time.

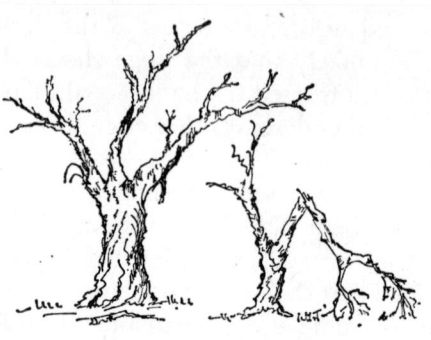

CHAPTER 10

The Ostriches

WHAT with one thing and another, the Field Force was once again greatly under strength, and their morale was low following the shameful poaching incident and the death of the Ranger during the course of the investigations. A Refresher Course was therefore planned for them, and steps were taken to recruit more tribesmen from the Northern Frontier District who would undergo training and bring the Force up to strength once again. As before, these men were selected from the Turkana, Samburu, Boran and Orma tribes, proud warriors all, who were plastered in the traditional red ochre and adorned with ostrich feathers and beads.

Amongst members of the Turkana tribe, especially, anything to do with ostriches has very great value, be it feathers, skin, or pieces of shell, for these commodities form a vital part of a warrior's traditional dress and play an important part in the customs of the tribe. The ostrich egg shell is fashioned into beads for ornamental purposes; it is also used by witch-doctors in a surgical operation undertaken to relieve persistent headaches, when a portion of the patient's skull is lifted out,

127

and replaced by a piece of egg shell; firmly believed to be a most effective cure, albeit rather a drastic one!

The feathers are greatly coveted to adorn a warrior's elaborate headgear. The natural hair is first mixed with clay and moulded into a rigid *chignon* at the back of the head, and a wooden comb with a series of holes is secured to this to hold the ostrich feathers in much the same way as a flower holder in a vase. Many hours go into creating these elaborate tribal hairdos, and in fact such importance is attached to them that a warrior cannot sleep with his head on the ground for fear of disturbing the arrangement. Instead, he sleeps with his head raised off the ground by means of a wooden neck yoke, and in this way, although probably not the most comfortable way to sleep, his hairdo remains intact and requires only the minimum amount of repair. Several thin sticks stuck into the clay chignon ensure that he can scratch his head should the need arise, again without disturbing the creation too much.

The day our recruits arrived in Voi, the larger orphans, including the three half-grown ostriches, were all feeding peacefully below the offices as the lorry noisily swung into sight; its fearsome looking cargo all chanting and shouting excitedly on the back.

As though drawn by a magnet, every eye was instantly focused on our unsuspecting ostriches the moment they came into view. A noticeable hush was followed by a surge of tremendous excitement, and the next moment several warriors had leapt clear off the back of the moving vehicle, and without even hesitating, had begun to leg it in hot pursuit of our three astonished ostriches, who in turn lost no time in getting under way. The flabbergasted Sergeant in charge of the Field Force, who was decked out in all his regalia to receive the recruits, began to bellow instructions interspersed with abuse at the top of his voice in a futile attempt to restore some semblance of order and recall the unruly ostrich addicts! Finally, three rather dejected looking warriors were rounded up and roundly reprimanded by the irate Sergeant, but he might have spared his breath, for they didn't even understand a word of Swahili!

Three months later, however, having completed their training, these same men looked very different when they paraded smartly for the final inspection before going into the Field. Only one thing slightly distracted that line of disciplined Rangers, and even this would probably have passed unnoticed had we not been watching for it: the arrival of the three ostriches on the parade ground, who strolled in front of the column in a most tantalizing way, while every pair of eyes moved slowly with them, following every single movement avariciously.

The parade ground quite obviously held an intense fascination for the ostriches, and there was no doubt at all that they thoroughly enjoyed watching the Rangers drill. The sound of the Sergeant's raucous voice was the only signal necessary to send them hurrying to the scene. On many occasions they even appeared eager to emulate the Rangers, and would line up alongside the column, or walk slowly up and down the ranks just as though they were dignitaries inspecting a Guard of Honour! A passing visitor witnessing this unusual spectacle one morning, actually formed the impression that the ostriches were also undergoing training, and was heard to relate this rather incredulously to other guests at the Lodge.

I often wondered if the ostriches realized that when the Rangers were on the parade ground, a truce was enforced, for this was one of the few occasions when they could be sure of not being chased. They seemed to derive satisfaction from being able to tantalize their would-be pursuers in this way. There was no doubt that a certain sparseness in the ostriches' tail feathers could be detected since the arrival of the Turkana contingent.

As our ostriches grew older, it became apparent that we had two cock birds and one hen, as their drab adolescent feathers began to be replaced with the black adult plumage. The two cocks became known as Solomon and Grundy, while their sister simply became 'the hen'.

When in breeding plumage, the Somali cock ostrich is an extremely handsome bird. A bright red beak and shins stand out in sharp relief against the delicate blue neck and legs peculiar to this northern race of ostrich, as opposed to the red neck and legs of their southern Masai relatives. The feathers are predominantly jet black with others of pure white tipping the wings. Solomon and Grundy, although not yet looking their best, were nevertheless impressive birds. They had taken to wandering around independent of the other orphans, while their sister, who was not so bold, opted to remain close to the headquarters, where her favourite haunt was the workshop and garage, and where she seemed to thrive on a diet of nuts, bolts, nails and all sorts of other unlikely objects that caught her eye, suffering no apparent ill-effects. This taste for hardware caused the garage staff a good deal of inconvenience at times. Engrossed in an intricate task of piecing together the various components of an engine, his bifocal spectacles adjusted at just the right angle, Gordon Tweedie, the mechanic, usually failed to detect the ostrich's silent approach. Suddenly a vital piece of equipment would be spirited away right before his eyes, and swinging round, fumbling with his bifocals and wondering if he had seen correctly, he would find himself face to face with the hen, standing nonchalantly behind him, while the precious object appeared as a slight swelling travelling rapidly down her long slender neck! One day she even attempted to pluck Gordon's spectacles off his very nose, an insult which resulted in vigorous reprisals, with the ostrich being pursued round and round the workshop accompanied by a hail of abuse.

The hen, despite all this, was not unpopular around the garage, and many of the more junior hands found her presence a great convenience, for she provided an appropriate alibi for missing screws, bolts and other items.

'Where's that small bearing from the tractor?' Gordon would rage.

'I put it here yesterday; what the devil have you done with it?'

After a brief search, the garage hand had only to mutter, 'Shauri ya buni' and there wasn't much more that could be said, for this possibility was entirely feasible.

Buffalo are now very common

While the hen was extremely friendly and gentle, the same could not be said of her two brothers, who began to try and assert their authority as they became older, and ended up by becoming quite a problem. They took to chasing people, and so it was decided that the time had come when they must leave the fold and live a natural existence away from civilization. One day, all three ostriches were enticed into a travelling crate, and taken to a point beyond Aruba dam on the south bank of the Voi River, some twenty-five miles from the headquarters, where they were released and bid farewell. The probability of their being able to return to headquarters never even entered our heads, for they had been enclosed in the wooden crate during the journey and could have had no idea in which direction they were heading. But, to our utter amazement, four days later, who should stroll into the garage but the hen! It is a mystery indeed how she accomplished this feat, but we all felt that having found her way back, she should be allowed to stay, although Gordon looked slightly dubious.

The rains had broken in earnest a day or two after the ostriches had been left at Aruba, and for two months the Voi River raged as a torrent, filling the dam overnight, and racing over both spillways to form a shallow lake in the valley below. Although we did not realize it at the time, it was this that had delayed the return of Solomon and Grundy, who had found themselves cut off on the wrong side of the Voi River when it came down in flood. Nevertheless, they too made their way homewards, and just as we were congratulating ourselves on the fact that they, at least, had settled down in their new surroundings, and were unlikely to return so long after they had been released, a report was received from the Sisal Estate. Two extremely friendly cock ostriches had appeared, and were consorting with the labour! Soon after this, the Voi River subsided, and sure enough, home came the wanderers, looking even handsomer than when they had left, with bright breeding plumage complete with red shins and beaks, necks the texture of velvet of a most delicate shade of blue, and thighs to match. Strangely enough, the reception they received from their sister, who had quite obviously forgotten them, was anything but friendly. She flatly refused to have anything at all to do with either of them, and would not even tolerate them anywhere near the elephant stockade at night, where previously all three had slept quite happily together with the elephants. So, Solomon and Grundy had to face the nights outside the enclosure, not that this was any hardship, for they were quite used to being out, and it presented an opportunity to raid my garden unmolested.

Their efforts to ingratiate themselves back into favour with the hen were interesting to watch, for they had reached the stage when they were now beginning to take an interest in the fair sex, and indulged in the most elaborate courtship dances and displays, which sadly left the hen unmoved. They would crouch down, their wings outstretched to reveal the beauty of their feathers, and would swing their necks in a wide semi-circle; first to one side, then the other, almost sweeping the ground each time, and quivering and rustling their wing feathers vigorously all the while. Several minutes of this would be followed by a short stamping rush forwards towards the hen, (who, invariably, would be paying not the slightest interest), and the neck swinging would be repeated all over again with even more intensity. Sometimes they would emit a loud booming noise, blowing out their necks in the process, or spar with each other, or even display to the garden boy in the absence of the hen.

Finally the day came when Grundy, who had grown larger and more handsome than Solomon, gave vent to his pent up frustration by driving poor Solomon away from the headquarters. During one of these chases, he took a quick sideways swing at a labourer in passing, inflicting a nasty gash on the unfortunate man's leg, and this sealed Grundy's fate. David decided that once again, he must be deported.

Ostriches can be extremely dangerous, and can inflict very severe wounds, not by pecking their victim, as one might suppose, but by making use of the long claw on their feet. They raise their legs, which are immensely powerful, and with a quick chopping action, can literally tear a man open from top to toe. Because of this, the wisest course to adopt when attacked by one of these creatures, is simply to lie flat on your stomach on the ground, for when in this position, there is little the ostrich can do, other than stand on you!

This time there was no question of being able to entice Grundy into a crate, he was far too wise for that – having been caught out once – so David laid a noose on the ground with the rope lashed over a nearby tree. By placing corn in the centre of the circle, Grundy unsuspectingly stepped inside, and when both legs were within the noose, a quick tug on the rope had him buttoned up in two seconds flat. He was then trussed up securely and heaved rather ignominiously on to the back of the waiting lorry, to begin his exile, this time, further afield, across the Galana River at Lugard's Falls and beyond the Mapea Gap; a distance of forty miles from headquarters – persona non grata.

This left us with Solomon and the hen, and all was peaceful for a time, until Solomon suddenly realized that he was at the top of the pecking order, and he decided that now he too was in a position to throw his weight around. This took rather an unusual form, however, for he developed a dislike of hats, and would chase anyone who happened to be wearing a hat, until the offending hat either fell off or was hastily removed. The staff soon learnt that it was advisable to remove their hats before approaching the headquarters where Solomon, usually stationed at a strategic point, casually scrutinized each passer-by, and was quick to bring defaulters to heel.

'How extremely courteous all the staff are!' remarked more than one visitor, never suspecting the true reason behind their manners!

But, obviously, this state of affairs could not be allowed to continue, for we could not run the risk of Solomon injuring someone who was a bit tardy in the removal of his hat. Matters came to a head when he treed the foreman for several hours, whom he had happened to intercept on his way back from Voi, and who had been riding a bicycle at the time. Obviously,

Solomon's dislikes had now been extended to bicycles, for the foreman had been pedalling along, minding his own business, when he was pursued by the ostrich. Solomon, of course, had no difficulty in drawing alongside the unfortunate man, whose legs, although working like pistons, could not coax any more speed out of his dilapidated machine. In the end, he had no alternative but to dive off the bicycle, and make for the nearest tree with all haste, where he was marooned for several hours with Solomon parading arrogantly underneath. Finally, his frantic cries for help were heard, and some men came to the rescue.

Needless to say, the next day, Solomon followed in Grundy's footsteps, and was transported to Mapea, where he was re-united with his brother, and no doubt put back into his rightful place in the pecking order, for a passing vehicle saw the two of them together some weeks later.

The hen, sadly, met a tragic end some months later, when an exasperated labourer, whose cake of soap she had just devoured, picked up an iron bar which happened to be handy and flung it at her, shattering her thigh so badly that she had to be put down.

CHAPTER II

1963—The Year of Independence

I T was evident by now that Kenya was well on the road to Independence from Britain, and that it was just a matter of time before an African Government took control of the affairs of the country. Many people, with events in the neighbouring Congo fresh in their minds, were leaving the country to start again elsewhere; a few decided to risk it and stay on, but most people were slightly apprehensive about the transition to self-government, and wondered whether it could be accomplished peacefully.

As far as the future of the National Parks was concerned, some people feared that with Independence, the land-hungry tribes that surrounded these areas would be permitted to walk in and do as they pleased, and several political speeches made by some of the African politicians seemed to support this disturbing conjecture. It seemed that yet another threat was to face Tsavo East and, added to our anxiety for the future of the Park, were doubts of a personal nature, with the knowledge that there would now be no long term future for David with the National Parks service, for the African leaders had made it plain that all posts held by 'foreigners' would be Africanized in due course.

Summing up the position one had to face the fact that one was probably a fool to stay on and risk so much, and that it would be wiser to move on while one could and begin a new life somewhere else before it was too late. For, while all Government servants, including our counterparts serving with the Game Department, had been promised generous compensation from the British Government for loss of career, and were to enjoy a pension as well, the same benefits were not extended to those serving with the National Parks, which was only a quasi-Government body and not a true Government department. When the Parks had been created, in order to insulate them from political pressures, it had been considered advisable that they be administered by a Board of Trustees, who, although selected by the Government, kept the interests of the Parks paramount, and so acted as a buffer against political expediency. And so, it came about, that the very safeguards designed to protect the Parks themselves, penalized those few dedicated men who had created them. Officially, they were not civil servants, in the true sense of the word, and therefore they would not be eligible for any of the benefits extended to Government officials at the time of Independence.

Suddenly the future looked very bleak, and a poaching incident at this time didn't make us feel any more confident either. A patrol of the Field Force had encountered a gang of poachers hunting with dogs in the Park near Maktau, and had succeeded in arresting one man in an ambush. The other two offenders ran off, pursued by the Rangers. One Ranger, however, had been detailed to stay behind and guard the prisoner.

Suddenly, this poacher decided to attempt an escape, and attacked the Ranger with a large club he happened to have been carrying. The Ranger successfully parried the blow with his rifle, shattering the club into a hundred pieces, whereupon the poacher seized the rifle and tried to wrest it from him. During the ensuing struggle, the Ranger, who felt himself being overpowered, pulled the trigger and shot his assailant dead.

This incident happened near a railway encampment and very soon a large hostile crowd of railway workers, armed with sticks and metal bars, had surrounded the Ranger and the body, in a high state of excitement, hurling abuse and advancing threateningly. By firing shots into the air the Ranger managed to keep the crowd at bay and attract the attention of his comrades at the same time, who rushed to his rescue, and succeeded in driving the crowd back.

An urgent radio message brought the news to David, who, accompanied by a police officer, hurried to the scene. By this time the crowd had worked themselves up into a state of near hysteria, and as the car drove up they surged towards it in an extremely ugly mood. For a minute or two the situation looked like getting out of hand, but then there was a hush as David and the policeman alighted from the vehicle and, deliberately ignoring the crowd, strode past them to where the Rangers were standing in a huddle. This had a sobering effect on the mob, but one man, obviously the ringleader, stood there, sinking his teeth into his own arm and shouting:

'Oh, why can't I call up enough courage to kill these *Wazungu*! I am as a chicken! Wait until *Uhuru* – we will kill them all!'

For several minutes he continued to shout in this vein, but no one lifted a hand; instead, they slowly dispersed.

The police carried out their investigations, and much to everyone's dismay, a charge of murder was levelled against the Ranger responsible for having killed the poacher, and he was taken into custody.

This not only had a very demoralizing effect on the Field Force, but on us as well, and no stone was left unturned to secure the man's release. Finally, the charge was reduced to one of manslaughter, which in itself was serious enough in view of the circumstances. A date was fixed for the hearing in the Supreme Court in Mombasa, and much to everyone's immense relief, at the trial, the Ranger was more than vindicated. The Judge ruled that he had acted in self-defence, and commended him for carrying out his duties under very difficult circum-

stances, acquitting him from the charge. Nevertheless this incident, and particularly the remarks of the ringleader at the
railway encampment, left me, for one, wondering just what
Uhuru had in store for us.

My apprehension on this score was doubled when I discovered that I was going to have a baby! It couldn't have come
at a more inconvenient time, when not only was the country
in a state of flux, but our own future was so precarious. And
it couldn't have come as a more serious jolt for David!

'My God,' he groaned. 'This is the last straw.'

Nerves were frayed, and his lack of enthusiasm made me
bristle defensively.

'Anyone would think it was you who had to have it!' I
retorted. 'How about consoling me for a change! It's bad
enough having umpteen game orphans and you to have to
run around, let alone a squawling infant at this stage of the
game!

'And it's thanks to you entirely that I'm in this terrible
predicament anyway!' I added as an afterthought.

Poor David looked completely bewildered, obviously unaccustomed to a woman's logic.

'I'm sorry,' he said quickly, and carried on to try and convince me that perhaps it wasn't such a disaster after all, and
that just the initial shock of the glad tidings had resulted in his
unenthusiastic reaction, but that really he was delighted at the
news! I must say, he didn't look it! However, by the time our
baby was due, we had both got used to the idea, and were
beginning to look forward to its arrival with great eagerness,
until finally the great day came, and off I went into hospital
in Nairobi.

David paced the floor all night, imagining the worst, smoking
endless cigarettes, and displaying all the trepidation of the
conventional expectant father. By the morning, he was apparently a bundle of nerves, and literally leapt into the air
when the telephone suddenly rang, feeling more nervous than
when confronted by an enraged elephant!

'Mr Sheldrick?' a voice enquired.

'Speaking.'

'You have a daughter.'

'My wife, is she all right?'

'Yes, of course. She would like to see you.'

According to an account by my sister, the relief on David's face had been overwhelming, and the sudden easing of the almost unbearable tension in her household must have been welcome. He scrambled hurriedly into the car, and drove immediately to the hospital, I think, fully expecting to find me half dead. Certainly he looked extremely surprised when he peered hesitantly round the door and found me sipping steaming tea with the midwife instead. The incredulous look on his face made us both burst out laughing.

'I'd better warn you,' I said, 'our daughter's no beauty.' The nurse escorted David to the nursery, and handed him a bundle from which peered a tiny wrinkled face. David held it as though it would bite! Angela certainly couldn't claim beauty, for her nose was off centre, and her brow puckered in a frown, while two deep blue eyes stared solemnly into space. However, the nurse assured us that all these defects in her appearance were only of a temporary nature and that within a few days she would be in better shape to receive visitors.

In the months that followed, Angela not only thrived, but showed signs of becoming a dominant force in the household, and undoubtedly had her father well in hand.

Independence day came, and passed – peacefully contrary to the predictions of many people. Nairobi had been a hive of intense activity preparing for this auspicious occasion, and all other matters, including of course the elephant problem, sank into insignificance beside the importance of this chapter in the history of Kenya. The Board of Trustees responsible for the National Parks had been dissolved, and a new Board was to be appointed by the African Government, but this was likely to take some time.

In the meantime, an effort was made to improve the distribution of water throughout the Park, for it was obvious that here lay the bottleneck as regards numbers of the smaller animals. There had undisputedly been a very marked increase in many species since the Park came into being. Buffalo were no longer rarely seen, but were becoming quite a common spectacle in large herds. Zebra, too, had responded to the appearance of a more open habitat, and now occurred in far larger herds, while waterbuck, impala, eland and oryx all had multiplied substantially. Nevertheless, their numbers remained restricted to the carrying capacity of the dry weather ranges; those few areas within access of the permanent water which the animals had to fall back on when the natural rain pools dried up. They would be confined to these areas for six months of the year, which left large tracts of good grazing unproductive most of the time, and the grass became rank and unpalatable for the smaller species.

In areas of impeded drainage, *hafirs,* or catchment dams, were established, to collect and hold the rainwater for longer periods, delaying the time when the game would have to fall back on to the permanent water. Boreholes were attempted, but only two out of the many that were sunk proved successful and yielded water in sufficient quantities to pump. In view of this, rather than risk the precious limited funds left in the kitty for water development on projects which were something of a gamble, it was felt wiser to concentrate on the safer surface supplies; more hafirs and deepening existing natural pans.

By now, poaching in the Park had practically died out. Only very seldom was anyone brought in, and then usually for only minor offences, like collecting honey or trespassing in the Park. The Field Force maintained their vigilance, and constant patrolling continued, but now most of their excitement was derived from encounters with wild animals as opposed to poachers.

There was one occasion when a Ranger stumbled upon a sleeping cow rhino and her calf while patrolling in long grass. The rhino was on its feet in a flash, and before the Ranger knew what was happening, he found himself tossed high into the air, before the cow rhino made off. Unfortunately, its calf, which was following along behind, decided to stop when it got to the place where the Ranger had landed and where he was lying, stunned and badly bruised. All his frantic attempts to chase it off failed, and soon, to his horror, it began to call for its mother! She promptly stopped dead, listened briefly, and then returned at the trot to see what was upsetting her baby. The unlucky Ranger was promptly tossed a second time, and landed in the middle of a wait-a-bit thornbush, making him even more uncomfortable, while mother and baby proceeded on their way. Miraculously, apart from extensive lacerations caused by the thornbush, a sprained knee and bad bruises – and, of course, shock, the Ranger suffered no serious injuries.

Another man had a lucky escape when a rhino hooked him by the belt and carried him on the end of its horn for some distance, until the web belt snapped and he fell to the ground. By even greater good fortune, although the rhino lumbered right over the top of him, it failed to actually tread on him, and carried on its way. All the Ranger had to show for this gruelling lift was a grazed knee!

Once all the excitement of Independence had subsided, and the new Government had become operational, there came a welcome assurance that the Government was conscious of the value to the country of wildlife conservation, and had no intention of making any alterations to the boundaries of the Parks that had been defined prior to Independence. Mervyn Cowie had been right when, with great foresight, he had placed the emphasis on tourism in those early days, however irksome it appeared at the time, for by so doing the benefits of National Parks had been demonstrated and the potential of the tourist industry as a major source of revenue could be seen by the new African Government. There is no doubt that the Parks would not have endured for aesthetic reasons only for such luxuries would have been beyond the Government in a country where land hunger was so acute. Game had to pay its way in this modern day and age, which it more than did, not in gate takings, but in terms of foreign exchange and the substantial hidden revenue derived from all the facets of tourism. It was the game that drew tourists to the country in the first place, and having come they required hotels, clothing, transport, curios, petrol, and food, the provision of which provided employment for thousands of people. Already the tourist industry lay second to coffee, and one day could provide the mainstay for the country's economy. When one considered the acreage under wildlife, and the acreage set aside for ranching, and compared the revenue from both, no one could say that game did not pay its way, but apart from this, the Government recognized the prestige value of their wildlife, and the fact that in their Parks lay a national heritage which was the envy of the rest of the world.

Our future, and that of all the expatriate Wardens, held no security now, and there were repeated calls in the National Assembly for Africanization at all levels. We knew it was just a matter of time before we would be replaced. Against his better judgement, David could not bring himself to abandon what amounted to a life's work when there was as yet no one to take over and continue where he left off. What had been achieved had taken years of toil, making the result doubly satisfying. He, like the other Wardens, found himself held prisoner by a dedication that ignored all reason, and which placed the interests of the Park first.

I fully understood these sentiments, and indeed shared them to some extent, although I was plagued by doubts as to whether the decision had been a wise one. And yet, too many animal lives depended on our continued presence until there was someone competent to take over; our own orphans, quite apart from the wild inhabitants of the Park.

The first post to be Africanized was that of Director, followed soon after by the appointment of an African as Deputy Director. A new Board of Trustees, now consisting mostly of Africans, took over where the old Board had left off, and one of the headaches their predecessors had left them was the elephant problem.

Three and a half years had passed since the elephant problem had reached a climax during the drought, and David now found himself in an unenviable dilemma. Being faced with the possi-

Lesser kudu

bility of a desert, he had pressed for urgent measures to reduce the elephant population at the time of drought, but he now harboured serious doubts about the wisdom of such a step in view of the vegetational trends that had followed the floods. Instead of a desert, he had seen the amazing recovery of those formerly devastated areas, with large parts of former bush-country colonized by perennial grasses. He had watched the sudden mass regeneration of thousands of *Acacia elatior* trees along the watercourses and the appearance of *Acacia tortilis* in amongst the new grasslands, as well as an increased regeneration of the large evergreen, shady *Melia volkensii* in some areas, which was not subject to damage by elephants, and would provide much needed shade. He had witnessed the birth of streams and springs, and he had come to the conclusion that the whole question was a lot more involved than he had at first thought, with each species so delicately attuned to its niche in the environment and its role dictated by Nature, that to tamper with the dominant one must set in motion a chain reaction with far reaching repercussions to the ecosystem, the final outcome of which was impossible to assess. For elephant are the key to Tsavo. They are capable of changing the vegetational pattern of an area completely; of converting huge tracts of land from thick bush to open savannah. They also play a vital role in the provision of water. By puddling with their great feet the depressions that trap rainwater in the wet season, they seal the soil and make it impervious. They create new water-holes by rolling and bathing in these puddles, plastering large amounts of mud on their enormous bodies, which is carried away with them when they leave. They make sub-surface supplies accessible to other less well-endowed creatures by digging away the sand and tunnelling out holes with their trunks. By trampling they can even raise the water table in watercourses, causing underground water to flow on the surface.

They are the pathfinders of the bush. Their trails link all the main watering points to the feeding grounds, follow the best alignment over difficult terrain, lead the way over mountains and through gorges, across plains and through thickets. They are also Nature's 'gardeners'. They uproot certain trees, ring-bark some species, prune others, and plant yet others over a wide area, dispersing the seeds they have eaten far and wide in their droppings to germinate and grow elsewhere. They provide fertilizer, depositing in Tsavo approximately 1,000 tons of manure per day, which is carried beneath the soil by millions of dung beetles, and sometimes buried as deep as four feet. And thus the vegetation consumed by the elephant has not really been destroyed; it has been converted and re-turned to the habitat in another form, for only when the energy cycle is interrupted and something is removed from the habitat to which it rightly belongs, can one say that it has been destroyed. Take the baobab tree, for instance; a giant among trees often attaining a great age, which is sometimes attacked and felled by elephants. It is certainly distressing to see this happening, but if one goes deeper into the question, one will discover that the baobab contains over thirty trace elements and rare earths, and is particularly rich in calcium. While the tree is standing, it provides shade and a nesting place for birds, but only when it is actually felled are the trace elements released into the soil and made available to other plants, and, in turn, animals. It is not beyond the realms of possibility that this 'destructive' practice of elephant is simply only another of their roles in Nature.

Almost every animal in Tsavo is dependent on the elephant in some form or another; many even for their food. Were the elephant not there to break down branches, making the foliage accessible at a lower level in the dry season, the browsers would certainly be adversely affected. Were the elephant not there to trample and crop the grass when it becomes rank in the wet season, keeping it short and palatable for the grazing animals, they, too, would suffer. For until the numbers of plains animals can build up to a level that can keep the grass short, the elephant is necessary to bridge the gap, for rank grass is both unpalatable and unsuitable for most grazers, and cannot be utilized by them.

With all this very much in mind, it was reasonable to be

cautious about the danger of too hasty a decision to interfere with Nature at this stage. Another aspect of the problem raised food for thought. Was commiphora woodland the climax vegetation for this area, anyway? The fact that it inhibited the growth of grass and permitted run-off and erosion during periods of heavy rain seemed to promote doubts, for Nature does not normally operate in this way. Was the country always like this? Here again, there were clues which pointed to the fact that it must have been quite different in the past; the presence of numerous cairnlike ancient graves throughout the Park, typical of those of the pastoral nomadic Galla people to the north, who would hardly have occupied the area had it been under thick bush, for their cattle would have been unable to survive in such country due to the prevalence of tsetse fly. It is reasonable therefore to assume that the habitat must have been more open at one time, and in order to get some idea of when this was, David turned to the diaries of some of the early explorers; men like Sir Frederick Jackson, Selous, Lugard and Patterson, who passed this way long ago.

In 1891 a report by a party carrying out a preliminary survey of the railway described the country between Taru and Maungu as 'a dead level plain of dull red colour, covered with patches of grass and a dense growth of aloes, acacias and a few Sycamore'. Obviously a botanist could not have been included in the team, for what the sycamores could have been is something of a puzzle, but it serves to illustrate that the country must have been thickening up at this time, because a few years later, when construction work started on the railway, a further report refers to the heavy mortality amongst the railway transport – 63 camels, 120 mules, 579 bullocks and 774 donkeys dying of trypanosomiasis between 1897 and 1898. And yet, other areas must still have been comparatively open then, for Sir Frederick Jackson in 1897 wrote of 'the parklike country below Ndi . . . and the open country East of Ndara and North of Maungu'; both areas that were very dense commiphora woodland when the Park was proclaimed, and could by no stretch of the imagination be described as 'parklike' or 'open'. Selous

recorded that he had 'never seen eland more numerous anywhere, even in the best days of South Africa, than at Voi',
another pointer to a more open habitat at that time.

Even some of the mysterious springs must have been active
long ago, for Lord Lugard, looking across the Galana River
to the Yatta at Lugard's Falls in the year 1890, wrote in his
diary, 'Far away among the opposite hills, we can see a big
waterfall, which must be a tributary from the North, but it
is impossible to say . . .' Today, at this same place, a small
spring has appeared within recent times. Who knows, given
time, it might again cascade over the lip of the escarpment to
form a 'big waterfall', and be seen as such by people standing
in the self same spot nearly a century later.

In the light of all this evidence, we couldn't help but wonder
whether the gradual conversions from woodland to open
savannah and back again were not part of a natural and perfectly normal cycle, enacted in this part of the world, with the
aid of elephant, since time began. Was it possible to halt this
trend and keep the vegetation in a static condition, and was it
desirable to do so?

References to elephant in the old diaries were also enlighten-
ing, and seemed to lend support to the evidence of cyclic vegetational patterns, particularly when correlated to the records of
the habitat. The German explorer, Krapf, who travelled
through parts of Tsavo East in 1850, laments in his diary the
fact that the elephant were fast disappearing from the area,
but doesn't specify why this should be so. While he refers to
thick country, he also mentions the proximity of the Galla
tribesmen to the east in the area roughly where Aruba dam is
now situated, which would seem to point to the possibility
that the country might have been in the process of being converted to savannah at this stage. That it must have once been
grassland is almost certain, for the Galla graves remain as evidence of pastoral occupation.

In 1890, when Lord Lugard walked from the coast right
along the Galana River, the tracks of large numbers of elephant
were seen at a point near Sala, but no further mention of the

presence of elephant is made until he reaches the Tsavo/Athi junction area, when he records the spoor of one.

Jackson and Selous, whose references to the area point to a more open habitat in 1897, do not even mention elephant; nor does Patterson, a keen hunter who was involved in the construction of the Mombasa/Uganda railway, and who spent many months in Tsavo at the turn of the century. Likewise, famous elephant hunters like Newman, Karamoja Bell and Sutherland found it necessary to travel inland in search of their quarry, which they would hardly have done had elephants been obtainable near the coast.

Men who were stationed in this area during the First World War cannot recall the presence of elephant, and it is not until 1920 that they are in evidence once again, by which time the bush is known to have been fairly thick. That it was very thick when the Park came into being in 1948 cannot be disputed, and elephants were certainly present in large numbers at that time. It would seem that the pendulum reached its limit round about the early 1950's, and then began its backward swing, with the country being opened up very rapidly ever since.

But what about the elephant? Was a population crash in a drought year the next thing which would be triggered off by the pendulum's swing? Or, would the elephant move out of the Park *en masse* in search of better feeding grounds before this stage was reached? Not enough was known about the migrations of elephant to be able to predict which of these two alternatives would come to pass if the cycle ran its course. Legend depicts them as roaming over vast distances, and people have always been inclined to accept this belief as fact, but it had been our experience that it was impossible to lay down any hard and fast rules in this connection, for the movements of elephant appeared to be entirely seasonal with a direct bearing on rainfall and the availability of browse and water. As no two years followed the same pattern, so the annual movements differed from one year to the next.

David was now tortured by doubts; doubts about the wisdom of cropping; doubts about where all this was leading; doubts

about what would be in the best interests of the Park as a
whole; and doubts about which course to adopt. Scientific
opinion appeared to be divided on the issue. Some scientists
advocated cropping the elephant, but an equal number were
of the opinion that a clumsy attempt at culling might even
aggravate the situation rather than improve it, and included
in the latter were many men with considerable experience of
African conditions, all of whom stressed the need for caution
when dealing with an ecosystem so delicately balanced. Could
one treat the elephant in isolation when all species were inter-
locked to some extent, each interdependent on the other, each
with its own niche in the environment? What was good for
zebra, for instance, may not suit lesser kudu, and what was
good for the elephant may be detrimental to a good many
other creatures, the vegetation and even perhaps the soil.

But, apart from this, the only feasible means of cropping
elephant artificially was the elimination of entire family units,
but would this solve the problem? It had been demonstrated
from work in Uganda that elephant populations responded to
increased population density by a natural adjustment of the
interval between calving, and also by delayed maturity, but
would not the removal of substantial numbers tend to make
life easier for those that remained, who would respond to the
better living conditions by an increased rather than a reduced
breeding rate? And so, having embarked on a policy of cropping,
was not one then faced with the necessity to crop forever more
in order to avoid a population explosion? And was this really
desirable in a National Park, which, above all, should be a
sanctuary? Did the paying public really want to see lorries
piled high with raw meat on its way to refrigeration plants and
biltong lines, or experience the uneasy suspicion that the elephant
they photographed today may find itself in those same lorries
tomorrow? Would not all this destroy the peace of a wild place;
destroy the sanctity of natural and unspoilt surroundings and
result in a great deal of disturbance? Would not constant
harassing turn the elephant vicious and a danger to the public,
or drive them out of the Park into neighbouring cultivation?

But, above all, had it not taken many years to bring the poaching under control; many years for the neighbouring tribes to recognize the sanctity of the Park? How could we suddenly explain to those simple people that while they mustn't poach, we were permitted to crop?

But, there was one thing about which David had no doubts, and it was this that influenced his decision probably more than anything else – the fact that the Park he saw today was a far better place than the sea of scrub he had been faced with fifteen years ago. This, and his faith in Nature, decided the course he would adopt, for man can seldom, if ever, improve on Nature, and where so many views conflicted was it not better to err on the side of caution rather than risk precipitate action?

'Although elephant are adaptable, and where grass is available, it forms the bulk of their diet anyway, no doubt when they have completed their role and the habitat can no longer support such large numbers, then Nature will take a hand, and the solution that emerges will be in best interests of the Park as a whole, with the balance adjusted to suit current environmental conditions. I'll have to tell the Trustees,' he added.

'But, you can hardly get up and say all this now, having been responsible for starting the scare in the first place,' I replied, agitated by the dilemma in which David now found himself.

He held his head in his hands and was silent for a long time.

'I know,' he said wearily. 'What a mess I have made of the whole damned thing! That's what comes of only seeing half the picture, but even though I once advocated the cropping of elephant, I am now equally convinced that it would be a serious mistake, and I have my conscience to live with. I'll have to tell the Trustees,' he repeated.

At the next meeting of the Board, the elephant problem came up for discussion, and David delivered his bombshell to a hushed meeting. The majority of the Trustees were perplexed, and not a little surprised. One man remarked wryly, 'we were supposed to be discussing how we would crop the elephants, now it seems we find ourselves discussing instead whether we should!' But David explained that he considered it his duty as Warden of the Park to put his views before the Trustees, and to inform them of the circumstances that had resulted in this change of heart.

There followed lengthy deliberations, and the Trustees finally decided, in view of the confusion which seemed to cloud the issue, that a scientific team should be established as soon as possible to look into the problem in detail and report its findings back to the Board in due course.

At the time, we thought that the headache caused by the elephant problem was nearly over, and David certainly felt a lot easier having got it off his chest, but in fact it turned out to be only the beginning. In taking this turning David had embarked on a turbulent course which would be accompanied by a great deal of controversy over the years, so that it was often impossible to discern the wood from the trees. A rather lonely road, from which former friends would fall, but taken in the genuine belief that it would lead in the right direction.

If the aim of the Park was to encourage the greatest variety of animals in the largest possible numbers, then surely this must be the right way? And if scientific opinion could not agree, was it not wiser to let Nature take its course?

CHAPTER 12

More Orphans

During all this time I too was kept extremely busy, not only trying to keep track of Angela, now a mischievous toddler bent on suicide, but also having to tend and care for an influx of new orphans who arrived in a steady stream.

The first of these was Toby, a baby duiker, found lying near the main road by a passing visitor. A wound round the base of the neck indicated that he had either been caught in a wire snare, or held in a predator's jaws, the former being the more likely.

Soon after, two more ostrich chicks were deposited on our doorstep, which were christened Romulus and Remus by Jill, and put into the wire enclosure to join Toby. The introduction was amusing, for Toby's territoriality came to the fore, and he resented the intrusion. He put his head down and advanced on the two chicks, trying to look as threatening as possible, but this left the ostriches quite unmoved, and so after several further attempts, he decided to give up and lie down in the corner instead.

A few days were spent ignoring the ostriches completely, until he became more sociable and finally developed quite a strong attachment to them, enjoying pushing them around the enclosure, chewing their feathers, and generally bullying them at odd intervals.

As for the ostriches, we had to provide them with a foster mother in the person of the garden boy, who had to bend down and pretend to peck at the ground with his hand so as to look as much like an ostrich as possible and stimulate the chicks' appetite. Baby ostriches are extremely delicate when very young, and also selective in their diet, picking out small legumes and weeds. Unless they can be persuaded to feed avidly they very soon collapse and die. But Romulus and Remus thrived thanks to the garden boy and Jill also, who spent many hours patiently gathering delicacies and tying them to the fence, so that the ostriches could pluck off pieces when the garden boy was absent.

Toby thrived as well, and was an independent little buck who never left the vicinity of the garden, but who was so adept at tucking himself away and making full use of his camouflage, that finding him when it came to bedtime was not always easy, and sometimes impossible. On these occasions he slept out, and once the search party sent out the following morning found him lying in some thick grass some distance from the house, with a lion and eight wild dogs not far away, so he was extremely fortunate to have survived that night. However, in spite of all the hazards that accompanied freedom, he soon acquired a taste for it, and the time came when we could no longer entice him into his box at night. He would struggle and kick so violently when picked up that it became quite impossible to hold him, and so there was no alternative but to let him be.

This caused me quite a few sleepless nights, when I would suddenly wake, conscious of every sound outside, imagining that I had heard him call, and would venture out with the torch only to find him lying peacefully beneath one of his favourite bushes, for duiker are crepuscular, feeding mainly at dawn and at dusk, and spending most of the rest of the time lazily chewing the cud.

Once Toby was permitted complete freedom, a very marked change came over him. Where before he had been rather aloof and timid, he now became friendly and affectionate, and would race up to greet any member of the family as soon as they appeared in the morning, and join us all for tea on the lawn in the afternoons. Favourite titbits included most fruit, biscuits, raisins, grain, popcorn and the seeds of the flamboyant tree, taken out of their elongated pods by Angela, who offered them to him in the palm of her hand. As a special treat we would sometimes pick rose leaves for him from my two treasured rose bushes, and this he greatly enjoyed.

One night, while returning late from an outing in the Park, we came across a lion in the road just below the headquarters, and as it bounded off at our approach, a second suddenly appeared from the left some twenty yards further on, while just beyond, the Armoury Guard, posted for night duty was stationed, blissfully unaware of their presence yet no doubt being closely watched by unseen glowing eyes. Thanks to our opportune arrival he was warned and advised to keep indoors on this occasion.

The night brought very little sleep, for the lions persisted in roaring and grunting all round the house until dawn, and I fully expected to find that they had dined on poor Toby, who was at large in the garden. However, in the morning, there was immense relief all round when he was found strolling nonchalantly around the lawn, apparently unperturbed in spite of what must have been a very harrowing night.

Just as Toby was beginning to sprout horns, of which he seemed inordinately proud, another little orphan turned up; this time a baby dikdik, who had been spared in the nick of time from providing a meal for its African captor by the Asian Security Officer to the Voi Sisal Estate. As I gazed at this minute little antelope, no bigger than a kitten, I wondered how anyone could contemplate such sacrilege! It had large soft eyes, a long elongated nose that 'wiffled' from side to side, and a little tuft of hair on the top of its head.

This little newcomer was promptly named 'Dicky' by the delighted children, and at once latched on to the nosedropper 'teat' proffered, downing the milk in a series of quick pulls.

Finding the most suitable milk formula for baby animals always presents some difficulty, for different species thrive best on different mixtures. For instance, whereas rhino and zebra, who actually belong to the same order, do best on Lactogen, which is designed for human consumption and has a fairly high sugar content, antelope will scour if fed on this milk. They seem to do best on plain milk, either fresh or powdered, diluted slightly with water. It is really just a question of trial and error, and of course, common sense, to determine the correct formula, beginning with a very weak solution, and gradually reducing the amount of water until one discovers what the animal can tolerate without scouring. But the surest means of judging its progress is by keeping a close watch on the animal itself. If its coat is sleek and shiny, its appetite good, and if it will play, then you can be confident that all is well.

The little dikdik was welcomed eagerly by Toby, who proceeded to lick him all over and was content to let him snuggle alongside when he slept. Dicky, on the other hand, whenever he felt the pangs of hunger, looked to Toby for food, butting him in the stomach vigorously as he would his own mother, but to Toby, this action was misinterpreted as a desire to play and he would respond accordingly with poor Dicky becoming more and more frantic. Dicky also enjoyed a game of tip with Angela, darting this way and that, jinking sideways at the last moment, and stotting on stiff legs round and round provocatively. He knew his name and unlike Toby, was relatively obedient, coming when called. He was an enchanting little pet, but alas not for long, for when he was six months old, we noticed that his coat had grown dull and his hair started to fall out in places. He lacked his habitual exuberance, having a rather hunched look about him, which, unfortunately, I attributed to the colder than usual weather we had been experiencing, for there was no sign of any loss of appetite on this occasion. But, all the while, Dicky must have been a lot more ill than we suspected, and by the time we realized this fact, it proved too late. One morning we found him in a state of complete collapse, his body as cold as ice, shivering pathetically, while his eyes and nose were congested, and his breathing rapid and forced. Overnight he had been reduced to a pathetic little creature in whom the spark of life barely flickered, and the expression of misery in those large brown eyes that had once been so full of animation, was heartrending to say the least. I couldn't help the tears welling in my eyes as I cuddled him in my arms, but there was obviously no time to lose, so I settled him comfortably in his box and went to fetch David to see what could be done.

'I think it's pneumonia,' said David after examining Dicky. 'But, I wish we had discovered it earlier,' he added.

A massive dose of penicillin was the best we could do, and as the needle punctured the skin, poor little Dicky gave a pathetic cry like a bird, and looked at us with puzzled eyes. How I wish we could have explained that what we were doing was not designed to torture him further but to help him! We carried him gently into the warmth of the kitchen, and rubbed his limbs to try and coax some life back into them. For several hours I stayed with him, encouraging him with soft words, and was rewarded by a sudden improvement in his condition, for he managed to struggle to his feet. I laid him down in a sheltered corner of the enclosure, urged him to take a little milk and left him to rest feeling a lot more optimistic. But the next day he suffered a relapse, and we didn't expect him to see another dawn. Throughout that night I kept getting up to peer into the box by the kitchen stove, and in the morning was encouraged when he managed to lift his head and even stagger a few paces, showing some interest in Toby who was pacing outside the fence. I opened the gate and allowed Toby to come in, and it was touching to witness his obvious concern for Dicky, for he walked round and round him, licking his face and nuzzling him gently, while Dicky responded, quite obviously very glad to see his friend. But this was to be a last farewell, for only half an hour later Dicky's body was cold and still, and his eyes glazed in death, while Toby stood dejectedly by. I knelt down to give poor little Dicky one last caress, and couldn't help the tears pouring down my cheeks. How fond one becomes of these little wild orphans reared from infancy: how different one feels about an animal when one can interpret its every gesture and expression, when by recognizing its individuality, it is raised above being 'just an animal'.

How arrogant some people are, particularly those pseudo-intellectuals who, sheltering behind the guise of science, denounce what they call 'emotionalism', and attempt to bring to ridicule those who are so misguided as to admit to a love of animals. Did they but know it, by so doing they admit their ignorance, for they could learn a lot more by being 'emotional' themselves where animals are concerned. After all, it was Dr Loren Eiseley, in his book *The Unexpected Universe*, who pointed out that the human soul 'craves that empathy clinging between man and beast, that nagging shadow of remembrance which . . . asserts our unity with life and does more. Paradoxically, it establishes in the end our own humanity. One does not meet oneself until one catches the reflection from an eye other than human.'

I don't know where David buried Dicky, and I have never felt the desire to ask, preferring to remember him as he was, a delightful little companion, who had captured our affections so utterly in the short time we had known him. We were to be lucky enough to get to know another three of his kind in the years that followed, each one an individual in its own right, but all equally endearing and intelligent, and all of whom contributed to making dikdik head the list of my favourite animals.

Toby, sadly, was another of our orphans who met a tragic end, but he led a happy and free life for two years first. We were able to save his life once when he was being relentlessly pursued one night by four jackal, and I was awoken by what sounded exactly like a baby crying. I listened for some moments, thinking that it must surely be a bird, but there was a breathlessness and urgency about the sound which made me decide to go and investigate, so I roused David. As soon as we were outside, there was no doubt that the sound was a distress call, but we still didn't connect it with Toby, for we had never heard him make any noise like that before. The next moment, to our horror, by the faint light of the torch we picked out an exhausted Toby in desperate straits, four jackal literally on his tail. Shouting frantically, David dashed forward to try and shake the pursuers off, but although this distracted them for a few moments, allowing Toby to gain some ground, they weren't going to be deprived of their quarry so easily, and soon took up the chase again. We then leapt into the Land-Rover and roared across the plain below the house in hot pursuit, bouncing and lurching over hidden stumps and potholes, until, finally, we succeeded in diverting the jackal and managed to drive them off by repeatedly steering them from the trail. We hoped that this would make it possible for our little buck to make good his escape.

The next morning there was no sign of Toby, and the day passed without him returning. We were beginning to fear that he must have fallen prey to the jackal after all, when to our great joy, he limped up to the front steps that afternoon, battered, scratched and extremely weary, but nonetheless whole and very glad to see us.

But the next time he was not so fortunate. Again, I awoke with a start, certain that I had heard him call, but the sound was not repeated, and I thought I must have been dreaming. Nevertheless, I went out and investigated the garden by torch-light. The night was quiet and very still, but nothing seemed amiss, so I returned to bed, sure that I had been mistaken. But, it had not been a dream, for the next morning we found poor Toby mortally wounded, lying beneath the hedge. Deep incisions in his throat and back had obviously been caused by the teeth and claws of either a caracal or a serval cat, which must have been disturbed when I appeared the night before. But what distressed me most, was the fact that I had passed within a few feet of where Toby had been lying, and had failed to see him. If only I had found him then, and been able to dress those deadly bites immediately, we might have been able to save him yet again. Now it was too late, and he died as we carried him into the house. We had had him for two years, and it was difficult to accept that he had gone forever. Once again, David buried a little buck, and when we asked where, he said 'Next to Dicky, of course'. We hoped that they had been reunited in a special place devoid of all predators, where the grass was green, and the streams crystal clear, set aside by a merciful Creator for the souls of the more gentle creatures of this earth only, and Angela is convinced of the existence of such a place. I hope she is right.

Toby was followed soon after by several other smaller additions, one of which was Chirpy, a little tree squirrel born in a fuse box at Mombasa and retrieved by a passerby, having been dislodged by the maintenance gang of the Power and Lighting Company. He was delivered to us in a bird cage, and included in the deal, with apologies, was a stout glove! Very necessary this was, too, for he had needle-like teeth and was quick to make use of them as well, latching onto a finger like a bull terrier. I wondered what I had let myself in for when I had promised him a home, and as things turned out, these doubts were amply justified!

The sight of this little creature confined so closely worried me, for I hate seeing wild animals restricted, and by the end of two days I could bear it no more and opened the cage door. Out he came, leaping over the verandah wall, through the lounge, up the curtain and out onto the back verandah with characteristic quick, jerky movements, his tail looking for all the world like a bottle brush. Having carried out a thorough investigation of his surroundings, he settled on the back verandah as a base, for here he could leap along the rafters, squeeze into the roof, or peer through the connecting windows into both the lounge and dining-room, as well as keep an eye on any activity in the kitchen. His food, nuts, bread, raisins and milk, was placed on the verandah wall, usually by me, and so after some time he developed a very strong attachment to me, which unfortunately was not extended to any other member of the family whom he obviously regarded as dangerous rivals for my attention, to be bitten whenever possible. I soon found that I could even dispense with the protective glove and handle him with impunity, picking him up and carrying him around on my shoulder. His devotion was really very touching. No matter where I was in the house, I had only to look at the nearest window, and there was his little face peeping in at me solemnly. If I moved into the bedroom, he would race round the house and appear at one of the bedroom windows; if I were in the kitchen, he would be there too! No matter which room I happened to be in, he would be at the appropriate window, watching my every movement, until David, who, incidentally, didn't share my sentiments where Chirpy was concerned, being the main target for his jealousy, said,

'That damned animal of yours is getting on my nerves! We never seem to be able to get away from it!'

Naturally, I defended my pet vehemently, and as though to get even with David for having made that uncharitable remark, Chirpy promptly darted out the moment he opened the back door, and fastened his teeth into David's big toe. An agonized howl followed by a string of oaths left us all in no doubt as to what had happened, and soon afterwards David came hobbling in, two tell-tale pin pricks of blood oozing from his toe, while from the nearest window, the culprit peered in maddeningly. Shortly afterwards, a resounding crash of breaking crockery came from the direction of the kitchen, and glancing up, I noticed that the face at the window had momentarily disappeared, so I had no difficulty in picturing what had happened. Back in the kitchen, my suspicions were confirmed, for I found the cook retrieving the broken bits, amidst much tongue clicking and muttering, and baleful looks directed at his toe, which bore much the same marks as that of David!

It wasn't very long before Chirpy had reduced the household to a state of jitters, for he persisted in the unfortunate habit of darting out of the most unlikely places to sink his teeth into a victim's foot, and hop back to the safety of the rafters to avoid reprisals. Not a day went by without someone being attacked in this way: agonized howls from Angela, roars of rage from David, screams from unsuspecting visitors and yells from the staff told their own story. It seemed that only I was immune, but, alas, not for long, although I was affected in a different way. Chirpy took to gnawing the curtains, which to me was a very serious development, although the family were secretly delighted! However, the curtains were old and could do with being replaced, so I made some allowance for him to begin with, but when he switched his attention to the precious new candle-wick bedspread, chewing a large hole in the middle, even I conceded that his continued presence was not in the family's interests. So Chirpy was caught, returned to the fastness of the birdcage, and expelled to the Voi River forest, where we selected a suitable tamarind tree, placed his nuts in a strategic crevice within the trunk, and set him free. We left him leaping exuberantly amongst the higher branches, exploring his new home with the greatest of interest.

Thereafter, we visited Chirpy about twice a week armed with an offering of nuts and raisins and invariably he would emerge from the leafy heights when summoned to accept what we had brought for him.

Finally the rains broke, triggering off all the excitement and activity that go with them, and for the first time, Chirpy failed to keep the rendezvous. His appearances became spasmodic, and so in the end we bid him farewell, confident that he had become adapted to his wild way of life, and was no longer reliant on us for his food.

CHAPTER 13

Fires—A New Hazard

Now that the elephants had removed the commiphora bush from large areas in the southern portion of the Park, and grass had become established in its place, a new complication was encountered, hitherto unknown in Tsavo East – fires. Now these raged out of control, fed not only by the dry grass, but also by an abundance of dead trash littering the countryside, and fanned to great ferocity by strong dry weather winds. Before the elephants got to work, it had not been possible for a fire to get away in the Park; indeed, on many occasions, David had deliberately tried to carry out controlled burning in certain areas in an attempt to open up the bush and diversify the habitat. He had even made a flame-thrower for this purpose, but this, too, had made little impression, for the bush simply would not burn. Looking back on this, it seemed yet another irony that he should now find himself preoccupied with the task of trying to control fires that swept into the Park from outside, and raced over extensive areas leaving desolation in their wake.

The main source of trouble stemmed from the railway, which

formed the Park boundary in certain places and actually cut into the Park itself in others. A few of the fires originated from sparks spewed from the locomotives themselves, but the worst culprits were undoubtedly the railway employees, most of whom seemed to be incurable pyromaniacs. Those responsible for the maintenance of the line lived in *landhies* or encampments along-side it, and every dry season, grass fires originating from these landhies, born perhaps of only a few live embers tossed care-lessly from a brazier, ravaged huge areas of the Park and destroyed hundreds of acres of valuable grazing, burning some-times on a thirty mile front for weeks at a time before they could be extinguished.

The only means of combating a large bush fire lay in back burning from a road or river towards the advancing blaze, thereby confining the damage as much as possible to a given area, for the flames leapt twelve feet and more into the air, and the intense heat generated precluded any approach. The back burning was usually carried out at night when the wind had dropped, and all the resources of the Park were mobilized on such occasions. Men toiled in shifts until the fire had been controlled, beating the flames down where they could, and ensuring that the back burn did not also get out of hand. One of the chief dangers in this respect stemmed from the many elephant droppings, which, when alight, went on glowing slowly for a very long time. The wind had only to lift the light fibrous particles and puff them far and wide, when suddenly a spark would be fanned to a flame many yards away, and yet another blaze would spring up and gain momentum, raging, roaring and crackling forward at an incredible pace, while an acrid pall of smoke cast a grey blanket over the country for hundreds of miles, making the sun take on a dull red eclipse-like appearance, and the air a suffocating heavy smell.

Ahead of the flames, thousands of insects; grasshoppers, beetles, and anything blessed with an ability to fly, rose in a dark cloud to escape death, while the buzzards, hawks and many other birds swooped down on this feast. Small rodents, snakes and similar creatures not fleet enough to escape, sought refuge in pigholes and anthills to avoid being burnt, but even so, the death toll of young birds and baby animals must have been very high. Larger animals, with the exception of elephant and rhino, do not seem to be greatly worried by fire. In fact, eland, buffalo and oryx have been seen to run quickly through an advancing wall of flame and stand unconcernedly behind the fire in the charred powdery residue left behind by it, apparently unscathed.

The aeroplane was a God-send in the control of fires, for from above a very clear picture of the situation was possible, and men on the ground could be directed by radio. It was from the air on one such occasion that David witnessed extremely interesting behaviour on the part of a cock ostrich sitting on eggs. Despite the rapid approach of a very large fire, this ostrich refused to abandon the nest, and became completely engulfed by the flames. Circling overhead, David deduced that it must surely have been burnt alive, but when the fire passed on, to his utter astonishment, there was the cock ostrich, in solitary splendour, still on the eggs, looking rather conspicuous and exposed on a charred black plain, but apparently none the worse for the singeing. The flames must have swept over him so swiftly, that by keeping his head close to the ground, and fanning his feathers, he had managed to escape unharmed, and we couldn't help feeling that it was only fair that such devotion to duty be rewarded in this way.

The animals that seem to suffer most in a fire are rhino and elephant, particularly if they find themselves sandwiched between the main fire and a back burn. Their great size is a definite disadvantage, for they cannot avoid the fiercest heat, and the charred remains of whole groups of elephants have been found on several occasions. These were the lucky ones, for there have been unfortunate survivors who have suffered the most appalling agony from severe burns, great slabs of hide sloughing off leaving the inflamed flesh beneath exposed to become a wriggling mass of maggots. Every moment such an animal lives must be excruciatingly painful, and a timely bullet must bring about a merciful release.

Rhino are probably at an even greater disadvantage, for they lack the intelligence of elephants, and on many occasions actually try to fight a fire, snorting and puffing and making repeated futile onslaughts into the flames, instead of seeking a means of escape. Many rhino perish unnecessarily because of such stupidity.

The eastern boundary to the Park was another trouble spot, for charcoal burners were active in the adjacent areas, and were often responsible for starting the fires. And then, of course, an occasional cigarette tossed carelessly from a car by a tourist can also cause a great deal of destruction.

David was very worried about the effects of these uncontrolled fires in a low rainfall area like Tsavo East, for although he believed that some burning was undoubtedly beneficial and indeed necessary under certain circumstances, he also believed that while fire could be management's most important tool, it must never be its master. One had to be extremely cautious about the use of it, and have full control over both the timing of a burn, and the area to be burnt, for there was no doubt that fires over the areas of red laterite soil in low rainfall belts were extremely damaging and left the soil exposed and subject to erosion for many months. Quite apart from this, with such an unpredictable rainfall, one could not afford to chance the loss of too much grazing. For the same reason it was quite impractical to work to a set 'fire policy' dictated by outsiders, for in places like Tsavo, one had simply to 'feel' the way in this respect, and be guided by conditions prevailing at the time. Like a farmer, the Warden must use his own judgement and common sense, and if he has sufficient experience of his area, and a naturalist's intuition, he will know instinctively far better than anyone else what areas should be burnt and when.

It was very vital to the proper management of the Park that those fires originating outside the boundaries be prevented from sweeping in, and similarly that fires within the Park be contained to a relatively small block, minimizing both the loss of grazing and the damage to the habitat and vegetation. First of all, assistance in the way of funds for a firebreak to protect the vulnerable eastern boundary was sought and provided by various bodies concerned with wildlife conservation, which enabled seventy-five miles of firebreak to be constructed, consisting of two parallel cuts, fifty feet wide and a hundred yards apart, with the intervening strip kept burnt to provide a wide barrier. This proved extremely successful, but it also required a lot of maintenance, and burning a hundred yard strip for a distance of seventy-five miles is not as easy as it sounds, particularly as it was important to do this early on in the season before the grass became tinder dry and the fire risk great.

The next step was the re-aligning of the road system within the Park, so that the entire area could be divided into blocks, and the roads themselves serve a dual purpose and act as firebreaks as well. In this way, should a fire break out in a certain area, a back burn from one of the surrounding roads ensured that it was successfully confined to that one block. And while he was about it, David decided to 'elephant proof' all Tsavo East's road signs once and for all, which, up until now, had consisted of the usual cedar boards on rustic posts. The elephants seemed to enjoy uprooting these notices, or turning them around, to the confusion of the visiting public, and very often signboards lay at untidy, drunken angles. So, mini *kopjes* were erected at all the road junctions, fashioned from an agglomeration of natural boulders cemented together in such a way that they looked as natural as possible, with flat paving sandstones from Sala incorporated in the sides on which the lettering could be painted. Although the initial cost of these sign cairns was quite substantial, the maintenance thereafter was negligible. Moreover, they were more in keeping with the rugged scenery of Tsavo and were permanent.

The re-aligning of the road system also necessitated naming many new tourist circuits and places of interest that came about in the process; a popular pastime in which all the family took a hand. It had always been the practice to select Wakamba words for places in the northern area of the Park, and words from the Waliangulu tongue to name places south of the river. And so, with the help of ex-poachers and Rangers, who supplied the words we required, many new musical names with the English interpretation underneath adorned the new rock signs; Hadulo Basani (Hammerkop Pools); Dida Harea (The Place of Zebra); Kono Mojo (The Waterhole of Eland); Kuple Nek (A Pride of Lions) and Ashaka (Still Pools) to mention just a few – mysterious names that lent fascination to the places they described.

Here again, the aircraft proved its worth, and was almost as vital to road making as the tractors themselves. From the air the most suitable and interesting alignment for a new road could very easily be selected, and the road could be designed to skirt any waterholes in the area or follow the top of a ridge to allow the best possible view of the surrounding country. All the new roads in Tsavo East were plotted in this way. By flying overhead, David guided the tractors on the ground which demarcated the route indicated by cutting a rough trace through the bush to be cleared and shaped into a road later.

What a difference it would have made to the Park had the plane been available fifteen years ago, for those first roads could have been plotted with foresight instead of having to be hacked through the bush at random. It all seemed so simple now, but in the early days it had been like blundering around in the dark. There was no doubt that the aircraft had marked the turning point in the development and proper control of the Park, and had proved the most essential piece of equipment a Warden had at his disposal. It had telescoped a huge tract of country into a manageable unit, and looking back on it, David wondered how he had ever been able to work without it. He used it almost as frequently as his Land-Rover, and was equally at home in both modes of transport.

One day there was a report of a rogue elephant at Aruba Lodge. This crusty old bull was apparently in the habit of wading into the dam to circumvent the elephant-proof ditch surrounding the Lodge, and patrolling the Lodge grounds at night and at dawn, not only systematically munching his way through the many trees painstakingly planted to enhance the rather stark surroundings of the Lodge, but also terrorizing visitors and staff alike whenever they ventured out. As the kitchens were detached from the *bandas*, many visitors had either to run the gauntlet or go hungry. Obviously this situation had to be rectified, but thunderflashes did not have the desired effect, and only made the elephant more aggressive. Finally, it was decided that he would have to be shot, and so early one morning, David flew out to the Lodge to despatch the marauder. Circling overhead, he could see that things were already in full swing, for down went the laundry enclosure as the elephant burst clean through it, while figures sprinted for cover in all directions and anxious faces peered from windows. He hurriedly landed on the road just outside the perimeter, and seizing his .416 rifle, clambered out of the plane and ran to the scene. The elephant spotted him coming, trumpeted with rage and promptly charged. David, quite confident, took careful aim to make absolutely certain of the frontal brain shot, and pulled the trigger, but, to his horror, nothing happened. He swore softly, hastily ejected the misfire, reloaded and pulled the trigger again, with exactly the same result. By this time, the situation looked like getting out of hand, for the elephant coming like a ship in full sail, was almost on top of him. A lightning appraisal made him realize that there was time for only one more shot, and that if that failed, he would have to dive ignominiously into the elephant-proof ditch just behind. This time, by great good fortune, the bullet went off, and the elephant, borne forward by its own momentum, plunged and fell literally at David's feet in a cloud of dust.

After a short while, when everyone had satisfied themselves that the elephant was in fact dead, all the visitors who had been staying at the Lodge overnight poured out of their bandas, and hurried to the spot. One man, in a high state of excitement, exclaimed breathlessly,

'This is real Africa! I wake up in the morning, happen to glance outside to see if anything is around, and see an enraged elephant rampaging through a fence; then a bloke pitches up in an aeroplane, dashes up to the elephant, and waits until it is almost on top of him before downing it! It's unbelievable!'

The events of that morning, as viewed by this overseas visitor, must indeed have seemed very strange to say the least, and David didn't spoil the story by telling him how near he had been to taking to his heels! Instead, he came home, and disposed of all the remaining .416 ammunition, which was obviously very old, for a misfire under such circumstances could very easily have ended in disaster, and a repetition of such a situation could not be risked.

CHAPTER 14

Houdini and his Ilk

ONE of the strangest of all wild animals must surely be the black rhinoceros. An animal of contrasting characteristics, capable of extreme gentleness and affection, and yet so solitary and unsociable in the wild state; misunderstood, universally regarded as unpredictable and bad-tempered, the rhino ranks low in the animal kingdom in terms of popularity. And yet, when one has an opportunity of getting to know these animals intimately, one discovers not the vindictive, dangerous beast one expected, but instead a rather pathetic and endearing creature, gentle, sloppy and simple, and rather forlorn, whom Nature seems to have burdened with more than its fair share of handicaps and placed at a distinct disadvantage from the moment that it totters to its feet.

No animal could be more affectionate than our tame rhino, Rufus. He trusted and loved all humans implicitly, and would accept any offering with extreme gentleness, savouring the lingering flavour by sucking his tongue long after the last morsel had disappeared. No animal enjoyed the facilities provided for his comfort more than Rufus; he loved his stable, his mud wallow, his food and the sameness of his daily routine. He loved attention of any sort, and particularly having his tummy scratched, lying down with legs outstretched and eyes tight closed in bliss and contentment. He loved the children and allowed them to clamber all over him, or ride him like a horse. The more affection lavished on him, the happier he was, and yet, who would suspect this side to a rhino's nature from just viewing these animals in the wild state, where years of persecution have made them unpredictable and dangerous. Handicapped with extremely limited vision, they seem to have found that the best means of defence is attack, but more often than not, they don't even know what it is they are attacking.

Their natural anti-social behaviour extends towards each other, and that they should lead such a lonely and solitary existence in the wild state when basically they are capable of deep devotion and crave company has always puzzled me. For a rhino's greatest threat, apart from man, stems from his own kind. All Nature's rules of fair play seem to be waived where rhino are concerned: males will not only fight each other but will attack females and vice versa, females will battle together, males will kill young ones, and so on. One visitor to the Park witnessed a harrowing illustration of this peculiar trait, which he related to us later.

A bull and a cow rhino were engaged in mortal combat by the side of the road, and the odds on the result proving fatal for the unfortunate cow looked high for, not only was she a lot smaller than her opponent, but she had with her a small calf which she dared not leave unprotected. As the visitor watched appalled, the exhausted cow was being literally battered to death, going down time and again as the bull drove his horn into her body in a series of ferocious onslaughts. The cow was having great difficulty in regaining her feet, and was hindered even further by the presence of her terrified little calf which cowered by her side throughout. Try as he would, the visitor could not distract the attention of the bull: he shouted, thumped the sides of his car and blew the horn, but nothing made the slightest difference. In the end, he could bear it no longer, and drove off leaving them to it, reporting the matter at the gate as he left the Park. However, by the time the news filtered back to headquarters, and Rangers were sent out to investigate, no trace could be found of any of the parties involved.

It seems very odd behaviour that a male should seek to kill a female of the same species, particularly one with a small calf at foot, and the most likely reason probably lies in the territoriality of rhino. There is evidence to suggest that while a certain number of individuals will share a given area and tolerate each other's presence observing a kind of communal truce when they chance to meet, the sudden appearance of an outsider in the territory will be resented most strongly and will trigger off violent reprisals resulting in the intruder being either killed or hounded out. Presumably the unfortunate cow had trespassed that day and had paid the penalty.

Rhinos are rather difficult to rear in captivity, for they are delicate creatures and seem prone to go down with pneumonia or tick-borne diseases, particularly if there is any element of shock involved when they are brought in. Even a very small rhino will fight and charge his captors with all the aggression of his species, and will take quite a lot of overpowering. Several babies brought in during the drought had not survived the shock of capture, succumbing to a build-up of the parasites that are always present in the blood, but which under normal circumstances have no adverse effect on the animal.

Rufus had been newborn and without fear when he was found, and so, too, was Stub, the next rhino baby we acquired, who on arrival measured exactly 18 inches at the shoulder and tipped the scales at 66 lbs.

Stub was brought in by the Sisal Estate Security Officer, having been born on the Estate that morning and apparently rejected by her mother. Luckily the Sisal Estate Manager had arrived on the scene in time to save her life, and drive the mother off.

I could hardly believe my eyes when this miniature edition was lifted from the car and tottered a few paces on unsteady legs. Her skin was dark and smooth, her nose blunt and nude-looking with no trace of a horn, and her tongue a bright pink as it licked my legs in a quest for food. I offered her my finger, which she seized and sucked strongly, and so, thanking the Manager for his trouble, I hurried to the kitchen to set about mixing up a bottle.

Experience with Rufus had illustrated the fact that rhino seem to do best on Lactogen, but as he had been raised by the mechanic's wife, the actual details regarding the strength and quantities were not available. I decided to play safe and start with a very weak solution and work up slowly from there. By the time David came home, my new orphan was sleeping peacefully on a full tummy in the ironing-room.

'Guess what I've got,' I asked as he came in.

'An elephant,' said David.

'Wrong.'

'A mouse, then!'

'Wrong again!' I replied.

'Give up, then,' said David, accepting defeat.

'Come and see!' I said.

He was astonished when the ironing-room door was opened to reveal the minute rhino curled up in a corner. It looked almost foetal with its smooth, soft skin and its little toenails gleaming and clean.

'Has she got a name yet?' David enquired.

'Yes,' I replied. 'Stub seems just right!'

So, Stub she was, and from the beginning she proved rather a problem baby. In fact, we wondered if she may even have been slightly premature, for Rufus had weighed close on 80 lbs at birth. For two weeks I struggled day and night to keep her alive, juggling the consistency of the mixture, adding a little more of this and a little less of that, during which time the little rhino looked near to dying on more than one occasion. Pneumonia complicated the dietary troubles, but in spite of all this, Stub clung to life with a tenacious determination, until one day she was over the hump and took an interest in her new surroundings. We devised a sling to hoist the baby rhino

onto a spring balance every week, and my pride knew no bounds
when the scales recorded a steady weight gain of about a pound
a day. Very soon, Stub herself displayed signs of real health:
she began to play, running across the lawn, then stopping
abruptly, pivoting round several times in a series of little jerks,
tossing her head and snorting in true rhino style, and now and
then kicking out a hind leg before racing off again to repeat the
whole performance.

Things very often happen in threes, and sure enough, a
series of bellows from a car that drove up to the house a few
weeks later heralded the arrival of another orphan; this time a
buffalo calf, who had fallen into a ditch and been abandoned
by the herd. She was christened Lollipa, a name which was
descriptive of the ungainly way she ran. In no time at all she
and little Stub were inseparable companions.

We built a small enclosure and three stables just outside the
kitchen to house the nursery members of the orphan herd until
such time as they were old enough to graduate to the larger
stockades where Eleanor and the others slept. Stub used to feel
the cold at night, so we made her a jacket out of an old blanket
which was put on at bedtime to avoid the risk of a repetition
of the pneumonia that had so nearly cost her her life before.

Number three orphan soon followed, and turned out to be
a baby lesser kudu, found by Sisal Estate workers tucked away
in a bush, and rescued by the again timely arrival of the
Manager. Now the Nursery had its full complement, and the
finishing touches were put to each stable when the occupant's
name was painted above the door: Bobby, the kudu, at the
end, Stub in the middle and Lollipa the other side.

I now had my hands full in no mean way, for each baby
had a different milk formula. Feeding time required Angela's
co-operation, and she usually fed Stub the rhino, while I dealt
with boisterous Lollipa and Bobby. To facilitate matters,
Lollipa was taught to drink from a basin, but the difficulty
came in getting the basin of milk in position before a greedy
over-eager buffalo sent it flying! Bobby behaved with quiet
dignity at all times, and patiently awaited his turn.

The three little orphans had the run of the garden, as usual, so the door to their enclosure was ajar throughout the day, and they were free to come and go as they pleased. They never left the vicinity of the garden, though, unless accompanied by either David, myself or the children, and were content to stay near their stables, or amble around the house until the afternoon, when they got into the habit of expecting some attention and exercise in the form of a conducted walk. Three little faces would peer in at the front door, and we would be expected to drop whatever we happened to be doing at the time for this treat.

The whole family would set off down the road past the offices, Stub and Lollipa galloping ahead at top speed, to about-turn when they were almost out of sight and come tearing back again to us. After a while, when they had worked off their high spirits and excess energy, they would settle down to a more leisurely pace, strolling along beside us, pausing at intervals to investigate interesting smells along the roadside, or to sample a wild plant. Bobby usually followed some distance behind, keeping rather aloof and stepping with the exquisite grace of his kind until he had to bound aside to evade Stub's mock charges.

The children also took part in any races, but had to be adept at dodging at the appropriate moment when their four-legged friends were about to overtake them, for Stub especially had a rhino-like tendency to merely run over anyone who happened to be in the way. Many were the surprised and rather puzzled tourists, who, upon rounding a bend in the road, were greeted by the sight of a galloping baby buffalo, tail aloft, closely followed by a charging miniature rhino, some children and the more sedate adults, while a graceful kudu brought up the rear. Astonished faces would goggle unbelievingly from car windows at this unusual spectacle!

Lollipa became Jill's special favourite, who saw in her a means of satisfying the latest craze for horses. Jill spent many hours training Lollipa to a halter, alternatively coaxing and dragging her around the garden during these 'training' periods.

Lollipa did not object to being singled out for preferential treatment, and very soon responded to the lessons in halter leading. The main objective was to teach her to jump, and to this end miscellaneous obstacles were dragged onto the lawn and piled up to form the jumps. Then both Jill and Lollipa would treat the rest of the family, who were called upon to provide the audience and clap when necessary, to a show jumping display, the jumps being raised slightly after each clear round.

During these performances, much to Jill's intense irritation, serious disturbances occurred which usually stemmed from two sources, Stub and Angela! Angela, who complained that the jumps were too high, and always ended up biting the dust and making a noisy withdrawal; and Stub, who, not wishing to be left out of the game, simply ploughed straight through each jump, taking great delight in downing them one by one.

For many months, the daily walk continued to be the highlight of the orphans' day, and sometimes took them quite far afield, until one afternoon, when things could have ended in disaster. Fortunately, David happened to be present on this occasion, and Bobby had been left behind. Stub and Lollipa had galloped ahead as was customary, but just as they had spun round for the return leg, eight wild dogs suddenly leapt onto the road on either side of them in an attempt to surround them. Luckily, at the same time, the dogs became aware of our presence further up the road, and paused to look, slightly taken aback. Meanwhile, Stub and Lollipa, apparently quite unaware of the danger to which they were exposed, were busy racing back. Several of the dogs proceeded to give chase rather half-heartedly, keeping an eye on David and me, who by this time were shouting and running to the rescue as fast as we could. Fortunately, we succeeded in intercepting the orphans before the dogs were close enough to launch an attack, and managed to drive them off, but that day could have ended in tragedy. Had the wild dogs made a determined attempt on the life of either of the orphans, there would have been little chance of a reprieve, for they seldom leave the victim they have selected.

Following this incident, David decreed that any future walks must be restricted to the headquarters, for it was not safe for the orphans, or the children and me, to wander too far afield.

During this period, we had an opportunity of getting to know rhino even better when nine adults trapped by the Game Department in an area designed for human habitation were translocated to Tsavo East for release in safer surroundings.

It was felt advisable to keep the captured rhino in stockades for a period of three weeks after their arrival, in the hope that they would be more inclined to accept the area as their territory, and not wander out of the Park when released. As soon as the necessary preparations had been completed, the first rhino duly arrived in an enormous crate on the back of a lorry, and this one was followed soon after by the others. Nine was about as many as we could cope with at any one time, for they entailed a lot of looking after. Bush had to be cut to provide fodder during the day and at night, which in itself is no mean task for nine adult rhino. The stockades had to be cleaned out as best as possible, by means of a long forked stick inserted into the stockade to hook out old browse and débris, a process hampered greatly by the reaction of the occupants, of course, who continually attacked the stick. Water had to be laid on in each enclosure as well. All this tied up not only an entire Section of the Field Force, but a team of carpenters as well, for the stockades required constant repair following the onslaughts of the inmates, who were continually battering against the sides. Fortunately, after the first week or so, most of the rhino settled down and no longer objected so violently to the various activities designed to make their confinement more comfortable, but to begin with no sooner had the bush been thrown into an enclosure than it would promptly come flying out again, tossed aloft amidst terrific snorting and puffing, which sparked off a disturbance down the entire line as each rhino prepared to do battle. Likewise, the water barrels spent more time being heaved around like footballs than for the purpose for which they were intended. And also, as was to be expected, many of the more aggressive animals spent a lot of their time intimidating their immediate neighbours, who took up a defensive stance and were allowed very little peace. And so, the fairly orderly routine that eventually came about, made life very much easier for all concerned.

There was one rhino, the first to arrive in fact, whose initial attempts to escape from his stockade earned him the name Houdini – an enormous bull with massive horns, that stood out from all the others in size and was extremely ferocious. A good deal of the repair work had to be concentrated on his particular enclosure, and the logs surrounding it had to be well reinforced, for not only did he persist in pounding the sides with unbelievable force, but he also attempted to climb over them. Seeing Houdini as he first was, who would have believed that this bull rhino, shown a little kindness and attention, within three short weeks, would become as gentle and as tame as Rufus himself?

It was his size and ferocity that impressed the Rangers and prompted them to devote more of their time to him than to any of the others. Special delicacies were singled out for him, and after only a few days, he would accept these gently from an outstretched hand. Whenever he happened to be lying within easy reach of the side of the pen, the Rangers would scratch his tummy with a stick, and while this liberty was vigorously repelled at first, the time came when he would actually invite it, by lying down on his side and lifting his legs to make as much expanse of tummy 'scratchable' as possible. Later on, he even allowed his face to be rubbed, while a sloppy soft expression spread slowly over it, and his eyelids drooped contentedly. We thought all this remarkable in an adult wild animal, and especially a rhino, but what followed was even more so. The Rangers sat on the stockade sides with their legs dangling over, touching Houdini's back. Noticing that he didn't appear to object, but continued to chomp unconcernedly at his food, they then went a step further, and one Ranger cautiously lowered himself until he was actually sitting astride this enormous rhino. Even then, Houdini paid not the slightest attention, and permitted all the Rangers to take it in turns to perform this feat, allowing them to clamber on and off as they pleased. What other animal would become so docile within a short space of time? Within two weeks people could go into his enclosure with him, handle him, ride on him, stand on him and pet him with complete impunity, and this must surely go a long way towards dispelling the reputation with which all rhino seem to be labelled.

There was great excitement one day when one of the female rhino gave birth in the stockade, but, sadly, the baby suddenly died two days later. Although it had been seen to suckle, and everyone assumed that all was well, a post mortem revealed that there was no food at all in the calf's stomach. Presumably the mother must have been dry, and we wondered whether perhaps the drug used to immobilize her prior to her capture had any bearing on this.

Eventually the time came when the rhino were due for release. One by one the doors were opened and they were allowed to wander off into the Park, while we watched from the safety of the stockade sides as they hesitatingly explored their new surroundings, ambling slowly off until they were swallowed up by the bush. Some returned again at night, seemingly reluctant to leave the security of the stockades, but eventually all the inmates had gone, with the exception of only one – Houdini. Because of the strong attachment everyone felt for him, we had left him until the last.

'Can't we keep him with Rufus and Ruedi?' pleaded Angela, whose suggestion received widespread support from the Rangers. But David vetoed this, saying that Houdini was a wild rhino and had a right to complete freedom once again, and so the door to his enclosure was in turn opened. Houdini was in no hurry to leave, however, and had to be enticed out and the door firmly secured behind him to prevent his going back in again. Everyone assembled round him to bid him God-speed, patting his rump and rubbing his face, until the sun sank slowly below the horizon, and we left him browsing peacefully around the stockades.

Poor old Houdini! By befriending him we had unwittingly sealed his death warrant. He saw nothing amiss in wandering into the nearby thickly populated Teita Reserve, for he had learnt to trust and like humans. He must have been very puzzled when his innocent appearance put everyone to flight, and more puzzled yet when he lay down to rest in a small thicket near a village school and awoke to find himself surrounded by a screaming mob who proceeded to pelt him with rocks and stones.

Word travelled fast about the presence of this 'vicious' animal, which was labelled as a threat to the people of the district. They demanded his immediate death, and no amount of persuading could convince the people that left unmolested, this particular rhino had no desire to kill anyone, but simply wished to be friendly, and that if he were allowed to remain, he would probably move on and establish a home elsewhere. Finally, the allegations against Houdini reached such proportions that an order from high authority for his immediate despatch was received by the Game Department representative.

We did all in our power to try and save Houdini. David recced the area by air and on foot, hoping that it might be possible to immobilize him and transport him back into the Park, but several intervening deep ravines made the place inaccessible to vehicles. Anyway, by the time we got to him, we discovered that he was no longer the docile friendly rhino we had patted on the rump only a week ago, but instead an enraged and dangerous animal; the result of days of persecution and a betrayal of the trust in *homo sapiens* so foolishly taught to him by us. His predicament hung heavily on our consciences, for under normal circumstances, this rhino would automatically have shunned human habitation and would probably have survived. As it was, because he had lost his most valuable aid to survival – a fear of human beings – Houdini ended up by being shot.

The remaining rhino were seen periodically for some weeks

in the area where they had been released, but in spite of their three weeks' confinement, they too seemed prone to wander far afield, probably driven to do so in a quest for a piece of unoccupied territory, until they were finally lost in the enormity of Tsavo. All had been tagged prior to departure, and so were easily recognizable, and one cow, at any rate, seemed to have become re-established some thirty miles away on the Ndara plains, for we came across her fast asleep beside a typically red Tsavo anthill, and snuggled alongside her was a very small calf.

Meanwhile, our own little rhino, Stub, had grown apace, and the time came when it was decided that the three small orphans should join Eleanor and the larger animals, so that they could be taken further afield to forage. I had been dreading the introductions, especially as far as Stub was concerned, for both Rufus and Ruedi were now almost full grown and the formalities could be fraught with danger, if they decided to be anti-social. Eleanor, too, was hardly likely to relish the presence of yet another rhino, and Lollipa had to make the acquaintance of three other three-quarters grown buffalo and would also be heavily outnumbered. Bobby presented no problem, for he always kept aloof and was not unaccustomed to consorting with unlikely animals, having been raised with a rhino and a buffalo.

On the day selected, Eleanor and her brood were herded below the house, and Stub and Lollipa were invited to have a preview of who they were expected to meet from the safety of the lawn. Both stood stock still, staring at the larger orphans with great concentration, ears pricked and nostrils flared. Then Stub let out two or three terrific snorts, and this drew the attention of Rufus, who happened to be the nearest at the time. Ruedi had been deliberately kept at a distance for the time being, having been branded as the most likely source of trouble.

David then called Rufus, who advanced briskly anticipating something good to eat and, although Lollipa promptly bolted, Stub gallantly stood her ground. Standing beside her, I could detect a slight tremble in her legs, as she dropped her head, prepared for action, at the same time revealing the whites of her eyes to look as threatening as possible, confident that the little button on the end of her nose was an adequate weapon with which to defend herself from the immense opponent before her. Rufus, however, seemed so preoccupied by the thought of sweets that he failed to even see Stub, and strode straight up to David, towering over Stub like a giant. Stub hurriedly reversed a few paces, legs braced and manner defiant, but still Rufus ignored her completely. Obviously encouraged by this sign of weakness, she snorted boldly and managed a hesitant rush forwards, whereupon Rufus's tail shot erect and he pricked his ears as he gazed with interest at Stub, and everyone present held their breath. Then, he slowly lowered his head to sniff Stub gently for a few seconds before returning his attention once again to David, making it pretty plain that he considered her hardly worthy of notice. This was most encouraging, for us as well as Stub, who proceeded to prance around tossing her head and horning the ground in what looked like a victory dance, obviously satisfied that the enemy had been vanquished. It would, in fact, probably have been better for her had Rufus been a little sterner, for she grew bolder and more self-assured every moment, and finally decided to take on anyone else who was game. Before we realized what was happening, she had charged below the garden where the other orphans were grazing unsuspectingly, and where a kind of rodeo ensued, with Stub alternately chasing and being chased not only by the three large buffalo, but by the ostriches as well, while Eleanor also took a hand, trumpeting shrilly and rushing around with her ears out like two sails. This commotion attracted the attention of Ruedi, who proceeded to advance at a sinister trot, while I, at this stage, couldn't bring myself to watch a moment longer. Covering my face with my hands, I peered through my fingers just in time to see poor Stub sailing through the air, having been hoisted on the tip of Ruedi's horn. Fortunately she happened to land on the other side of a small bush, and while Ruedi was busy rounding the bush, David and the

Rangers managed to reach Stub first, and cover her ignominious retreat!

Back home we hastily inspected the damage. Miraculously, only Stub's dignity had been badly injured, for she bore just a few scratches, but she was trembling like a jelly and was a lot more subdued than before.

Meanwhile, Lollipa had taken over the ring, having been close behind Stub when her line of retreat had been cut by the three big buffalo. They then encircled her and began to advance on her steadily. This proved too much for Lollipa, who made a desperate bid to break through the ring, and in passing was also sent flying, bellowing loudly as she too sailed through the air.

It took some time to subdue Eleanor and her brood, who were all by now thoroughly roused, and were rushing around chasing the ostriches in all directions for want of anything better to charge. Finally, the orphan attendant successfully steered them further afield, leaving poor Lollipa limping home to join Stub, two very demoralized and dishevelled specimens in urgent need of reassurance!

'It's no good,' I lamented. 'Ruedi will end up by killing poor Stub.'

'We'll have to persevere,' said David. 'After all, we can hardly keep a two ton rhino in our front yard forever!'

And so, every day, renewed efforts were made to integrate the two, by now extremely reluctant, small orphans with the larger ones. This necessitated virtually dragging Lollipa out forcibly by the halter, and keeping her and Stub surrounded by Rangers to prevent them from doubling back to the house, while the other orphans were encouraged to feed nearby. Jill's lessons in halter leading had paid off, for without the halter it would have been well nigh impossible to persuade Lollipa to leave the garden after the unfortunate experience of the first introductions. As it was, in time Eleanor, Rufus, the large buffalo and the ostriches soon got bored with chasing them, although Ruedi, unfortunately, could never be entirely trusted, and for three months two Rangers had to accompany the orphans for the express purpose of protecting Stub. Even then Ruedi was no nearer becoming friendly, and persisted in taking a poke at Stub whenever an opportunity presented itself.

Finally we decided that he would have to go, for he was fully capable of taking good care of himself and was ready, we felt, to be rehabilitated to the wild state. Plans were set afoot to move him to the north bank of the Galana River, and release him near the tented camp, from where, it was hoped, he could establish a territory for himself without encroaching on anyone else's property. By this time, we knew enough about the behaviour of rhino to realize that the experiment was not going to be easy for Ruedi, but if any rhino could successfully revert, Ruedi could, and it was in his own interests that the attempt should be made. Although he was quite docile, like all of us there were odd occasions when he became irritated, or suffered the odd tantrum, and could unwittingly injure someone.

A stockade was prepared for him near the tented camp, and he was enticed into a travelling crate and transported to his new home by lorry, where he was kept confined for a week or so in order to become accustomed to his new surroundings. But, somehow, news of his arrival got out, for an old battle-scarred cow rhino accompanied by her half-grown calf turned up one night and broke into Ruedi's stockade, setting about him in earnest. The commotion was indescribable and roused the whole camp; terrified tourists peered nervously from their tents, flashing torches and trying to reassure one another, while the Manager, who surely deserved a medal, rushed out clad in his pyjamas and flung open the door of the stockade, enabling poor Ruedi to escape with the cow and calf in hot pursuit.

Rhinos make a variety of noises, and these, too, have extremes. They can 'mew' when calling in a 'wanting' way, they snort when alarmed, puff and squeal, and when thoroughly enraged, they roar; a deafening and terrifying sound not very often heard, but one which was heard by all the inmates of the tented camp that particular night. Gradually the pounding of feet, snorting, and roaring died in the distance, and people began to emerge from their tents to cluster around the Manager and hear the details of what had taken place, sipping strong coffee and listening with eyes as big as saucers!

No one really expected to see poor Ruedi alive again. David instructed an anti-poaching patrol which happened to be stationed in the area to try and trace him by following the spoor, but the tracks became obliterated by rain. Suddenly two days later, he hobbled painfully back into the camp, bruised and battered with a deep wound in his side. He staggered stiffly to the soothing waters of the river and lay down to bathe his wounds, a very sorry looking sight, and obviously not finding an independent life to his liking at all.

Ruedi's convalescence was long and painful, during which he spent a great deal of time actually lying in the river during the day. Nothing would induce him to sleep in his stockade though, and indeed he preferred to put the river between him and the cow at night. Unfortunately, but understandably, visitors to the camp lavished attention on him, which was the one thing we had hoped to avoid, for Ruedi was, after all, being trained to lead a natural life, and one of the most important aspects of the training was the need to sever his link with human beings. Instead, he was encouraged to come into the large bar tent, where he would rest his massive head on the counter, and open his mouth to receive any contributions; then shuffle out to sleep off the effects on the grass outside. Although motivated by kindness, this was, in fact, a great disservice to Ruedi, and only lessened his chances of successful rehabilitation.

In time his wounds healed well, and he took to venturing further afield to feed, although the aggressive cow and her offspring were still very much in evidence and continued to hound him for many months afterwards. Naturally, he sought the safety of the camp compound whenever she was after him, and many was the occasion that visitors had to dive for shelter when a thundering of feet gave warning that she was on the warpath! Fortunately, Ruedi, being younger and more agile, always managed to avoid actual contact, having learnt a lesson the hard way in the beginning.

The ostriches appeared eager to emulate the Rangers and would line up behind the column

Eleanor led the orphan herd

Eleanor would kneel down to help lift the calf

Sobo, standing beside the body of her mother, sprayed herself
continuously with water extracted from her stomach

While it had been fun and an added attraction to have an almost full-grown rhinoceros eating from one's hand, as usual the time came when Ruedi's presence in the camp was considered a hazard, and demands were made for his removal. Apparently when people failed to provide the goodies he expected, he was inclined to display signs of irritation, and the tour operators began to ponder the possible repercussions that would follow should one of their clients get skewered. After two tough years, Ruedi could qualify as being completely rehabilitated in every way but one. He had by then succeeded in establishing an uneasy peace with the cow, who had recognized his right to live in the same territory. We felt it would be a bit hard to subject him to the same painful process of establishing that right elsewhere. The experiment had failed because one vital aspect had been overlooked: we had failed to instil in him a terror of human beings.

We gave the matter of Ruedi's second move a lot of thought, and mindful of the difficulty he had experienced on the first occasion, we decided this time to send him to a privately owned game ranch north of Nairobi, where he would be spared the inevitable battle for a place, and could roam in peace, safe from poachers and the rigours of the wilds; the droughts, the floods, and the inhospitality of his own kind. So, once again, he was enticed into a travelling crate to face a long journey. He came via Voi on the first leg, and while the lorry was refuelling, I clambered aboard to have a look at him over the top of the crate, and bid him farewell yet again. He was certainly a magnificent specimen; enormous, strong and in fine condition. Gently he accepted an offering I had brought him and allowed me to rub the side of his face. I fervently hoped that this time he would be able to live in peace, and at the time of writing, it looks as though this hope will come to pass, for Ruedi has settled down well in his new surroundings, and seems to be privileged to the best of both worlds – freedom and the security provided by mankind.

CHAPTER 15

The Tsavo Research Project

THE elephant problem had always been bedevilled by sensational publicity, some of it not always accurate, and by 1966 it was again very much to the fore, with a flood of articles in the press urging action. Having had so much attention focused on the damage to the vegetation, most people had become conditioned to the belief that the only salvation for the Park lay in the destruction of large numbers of elephant. They came to Tsavo East with preconceived ideas and fixed opinions, and because of this their eyes rested automatically on the skeletons of dead trees littering the landscape, which only served to confirm what they had expected to find. The stands of perennial grasses went unnoticed, as did the better overall ground cover and the new springs and streams that had appeared. They overlooked the regeneration of *Acacia tortilis* replacing the fallen commiphora over wide areas and the marked increase in the numbers of plains animals; the wider field of vision they enjoyed and the greater variety of species to be seen. The passing of the commiphora was lamented out of all proportion without thought as to which of the two habitat types was the more productive from the overall faunal point of view. Every scarred baobab tree was a talking point, and every pile of bones subconsciously related to the 'destruction' of the vegetation. One reporter, certain that he had unearthed a 'scoop', photographed an accumulation of old elephant jaws which had been stock-piled over the years for ageing purposes, and captioned the photograph with a sensational story that elephants were dying in large numbers. Misleading publicity of this sort served only to confuse the issue even more, and soon everyone was in full cry. All sorts of 'experts' emerged from the woodwork to have their say, until the odd voice cautioning against a hasty approach to the matter was drowned in the general clamour for quick action. It was even hinted that several thousand elephant could be destroyed to help finance a research team for Tsavo; an extremely dangerous precedent, to say the least. The pressures from outside preventing an impartial assessment of the situation were a source of anxiety, for the wrong decision at this stage could be disastrous. It was therefore a relief when it was announced that the Ford Foundation had agreed to make available a grant to finance a research project within Tsavo, to study the problem so that guidance could be given on the future management of the Park with special emphasis on elephants.

Dr Richard Laws, who had been studying elephant populations in the Uganda National Parks, was chosen to head the new Tsavo Research Project, and a sample of 300 elephant was shot from the Tsavo population which was required to form the baseline for this study. This task was carried out by Ian Parker's firm, Wildlife Services Limited, and although David was opposed in principle to the idea of private enterprise undertaking work of this nature in a National Park, there was no denying that Wildlife Services had perfected the technique of cropping elephants in Uganda, and were in a position to carry out the work immediately. Furthermore, Ian was familiar with the parts of the anatomy required by the scientists for their studies, having worked with them in Uganda, so plans went ahead for the sample to be taken from the north bank of the Galana River at a place called Kowito, which was closed to the general public, and well away from the tourist routes. Very strict controls were enforced throughout the operation to try and avoid possible sensational publicity and the appearance of 'horror pictures' in the press, for the killing of elephant families is not a pleasant sight, but much valuable scientific data, which would have been difficult to obtain in any other way, was accumulated.

The actual method of cropping took advantage of the unit structure of the population and was gruesomely efficient. The doomed herd was driven by helicopters to the place selected for their annihilation, where men armed with semi-automatic weapons awaited them. The leader was shot first, and her sudden death reduced the others to a panic-stricken, bewildered mob, who clustered around her in complete confusion, utterly demoralized and not knowing what to do next. Any remaining adults were then selected, leaving the calves clambering over their mothers' bodies in pathetic terror, until they, too, died, and all that remained of that Family was an inert heap of carcases, the blood spilling out to form a sticky maroon pool before blending with the red Tsavo soil. But, they did not suffer for long – the entire operation seldom took more than three minutes.

It had been considered imperative at the time that the maximum use be made of the carcases, and the need to circumvent adverse criticism by ensuring that nothing was wasted was stressed before the operation began. The meat was therefore dried for sale, the feet converted to waste-paper containers or stools, the ears reappeared in the form of ladies' handbags or wallets, the hide cured to luxury leather; even the bones were removed, and, of course, the tusks sold to dealers in Mombasa.

But, in retrospect, had we, in fact, made the correct decision from the point of view of the environment? Had we not got our priorities snarled up somewhat, for in a National Park should it not be the habitat that qualifies for first consideration? Is not the importance to protect it written into the Ordinance?

'It shall be an offence . . . to remove from a National Park, any animal or vegetation whether alive or dead . . . to remove any object of geological, prehistoric, archaeological, historical or other scientific interest . . . to destroy or deface any object, whether animate or inanimate.'

We had lost sight of the fact that the loser was, in this case, the very thing about which we professed so much concern – the environment. Three hundred large mammals that would normally have gone towards enriching the soil had been lost to the habitat. For we know now that when an elephant dies its contribution does not end there. A complex and equally important chain of events is triggered off with the activity of billions of lesser organisms, each playing a vital role to ensure that nothing is lost to the ecosystem, but is simply recycled within it.

It is only when something is removed from the environment to which it should rightly belong that the term 'waste' becomes appropriate in the true sense, and there was no denying that the removal of large numbers of elephant from the Park was an interruption of the natural energy cycle, and as such detrimental to conservation. By making the 'maximum use' of those elephant, we had converted them to the material luxuries of our society, but they had been lost to the ecosystem: they had, in fact, been 'wasted'. We had erred in that we had mined the habitat, when as conservationists, we should have been mindful of the need to nurture it, for how could we replace those essential nutrients that were tied up in the bodies of those elephants, which should have been returned to the soil from whence they came, especially in a semi-arid area of low nutrient status like Tsavo East.

But then, in mitigation, hasn't man always had a regrettable tendency to manipulate the natural order of things to suit himself, and when it does, convince himself easily that what he has done was in the best interests of that which he has manipulated, conveniently overlooking any long term adverse effects of his actions? Cloaking greed under the guise of science, we piously preach present-day thinking and management in the light of modern research, but what, exactly, does this mean? We convince ourselves that it is the environment that stands to benefit from our interference, but then we deprive it of that which should be returned to it. Haven't we been guilty of the destruction of noble forests, the pollution of streams and rivers, lakes and seas; the disappearance of many species of birds and animals, and the impoverishment of the soil over huge areas? In our blindness, can we not recognize the signals that warn of the need to stop blindly accepting theories based on insufficient knowledge, and the need to call a halt to the mass exploitation of the environment? After all, haven't we consumed far too much of the earth's resources already; in fact, more in the last twenty-five years than the rest of man's occupation of this planet? Yet, we continue to heed that plausible patter about 'consumable natural resources' and 'maximum utilization of the habitat', including in this category the remnants of the world's wildlife which is struggling for survival. Fashionable to deride those who are emotional where animals are concerned; popular to denigrate those who marvel and take note of Nature's ways, enlightened to shrug off the very concept of Nature as something old fashioned and out-dated in the light of modern techniques, in the same way as it is considered clever to scoff at the concept of a God. Man has become so scientifically orientated that he fails to take note of the need to escape the trammels and stresses of modern living, a need which leads to a widespread longing for the natural and unsoiled, and a yearning for the solace of a wild place. With amazing arrogance we presume omniscience and an understanding of the complexities of Nature, and with amazing impertinence we firmly believe that we can better it. But, can we even begin to improve on the termite, or the honey bee? Perhaps, we have forgotten that we, ourselves, are just a part of Nature, an animal which seems to have taken the wrong turning, bent on total destruction. So prolific that we threaten the very survival of the earth, we stand in judgment of other populations guilty of the same crime, populations that will, however, bend to Nature's rules before all is lost. We, on the other hand, while priding ourselves in our superiority and ability to circumvent the inevitable, will be beaten by Nature in the end and will find that even we have to conform to her laws. And, if we are given the chance of hindsight when the time comes, we will then realize that we had not been quite so clever after all.

The Tsavo Research Project got off to a good start several months after the sample had been provided, when Dr Laws officially took up his appointment as Director of Research and came to live in Voi. Working with him was Dr Murray Watson of Serengeti fame, and John Goddard, whose particular facet was a study of the rhino population in Tsavo.

Dr Laws was soon immersed in a study of the population dynamics of the elephant of Tsavo, aerial surveys to establish movements, and censuses based on sampling techniques to ascertain the numbers involved.

A further sample from a completely different population, for comparative purposes, was then deemed necessary, and 300 more elephant were shot in the Mkomazi Game Reserve in Tanzania, again by Wildlife Services Limited. This was then followed by a request to shoot a further 2,700 elephants in Tsavo East, for the scientists had reason to believe the existence of ten discrete populations within the Park, and it would be necessary to remove a sample of 300 from each population.

The Trustees, who had to shoulder ultimate responsibility, had to give very careful consideration to a request of this nature where so many animals were involved. The aspect that influenced them most was the fact that there had until now been no Botanist on the research staff, and consequently no detailed study of the vegetation had been undertaken in the Park. Therefore, after lengthy deliberations, they concluded that the data available at the present time was incomplete, and that more information would be necessary, particularly on the botanical angle, before they could accede to this request. This triggered off differences of opinion between the Park authorities and Dr Laws, which resulted in the latter's resignation. It was the view of some people that undue prominence was being given to the elephant themselves, rather than to the ecosystem as a whole. Could the arid bushlands, for instance, support a large elephant population indefinitely without changes being inevitable? Was it desirable, or indeed even possible, to attempt to halt the present trends? Was it in the best interests to concentrate mainly on an elephant sanctuary and would the Park survive as such? If not, were the changes brought about by the elephant desirable or otherwise from the overall faunal point of view, and were they beneficial or not from the tourist angle? How many species stood to benefit from the changes in the environment and how many would suffer? Was the appearance of perennial grass cover a permanent feature of the vegetation, or only another transitional stage leading eventually to something else, and if so, what was likely to follow? It was hoped that a Botanist would be able to clarify some of these issues.

Dr Phil Glover, a man of wide African experience, who had been familiar with the Park and its problems for many years during his term of office in the Kenya Veterinary Department, and who had seen the vegetational changes from the beginning, was selected to fill the post. In his opinion, the new grasslands were more productive than the former bushcountry and the changes had certainly not been detrimental to the Park. This view, expressed by a qualified Botanist, was reassuring, and further encouragement came with the completion of the work on rhino by John Goddard, who concluded that the population appeared to be healthy at the present time, and totalled probably over 7,000, although he held certain reservations about the long term impact on the rhino by the presence of such a large elephant population.

Further appointments to the Tsavo Research Project followed, and included Dr Walter Leuthold, who was set the task of studying lesser kudu and gerenuk, and who worked also on buffalo and giraffe, and later on the elephants.

The Tsavo Research Project now became more closely affiliated with the University of Nairobi, and later with Oxford University, so that advantage could be taken of the facilities offered by those two establishments. A Research Co-ordinating Committee was also established to consider the many requests received from people wishing to carry out research in the Park, so that the contribution a specific project would make towards the particular problems facing management could be assessed before permission was granted. All applications to the Co-ordinating Committee were considered in this light in an attempt to ensure that priority was accorded to those projects that were likely to be of direct benefit to the Park, and also in order to try and avoid the common tendency of many scientists to make use of the facilities provided in a National Park simply to further their own ends, often working on studies of their own choice which have no bearing on the problems on hand.

Unless a definite chain of command is established at the beginning, and terms of reference clearly laid down, differences of opinion between the management and research workers in a Park are largely unavoidable, for one faction is essentially practical while the other is academic. A widely differing approach to a given set of circumstances can very often lead to misunderstanding at the best of times, but the breach is inclined to widen further when each becomes more deeply involved in his own field, and when the question of priorities comes into conflict. Each side understandably considers their particular work as more important, and resentment is likely to set in when opinions differ in this respect. One side usually feels that research in a National Park should be complementary to the management of the Park, and should be carried out unobtrusively and discreetly with due regard to the regulations safeguarding the sanctity of a Park, and with the minimum amount of disturbance. This attitude is frequently misconstrued as unnecessarily obstructive by ambitious young researchers, eager to produce quick results and make a name for themselves in the scientific world. To one side it is repugnant to see animals in a supposedly wild setting degraded with streamers and identi-fication tags, and sacrilege to deface the natural scene with con-crete markers and white paint, while to the other, this is an essential part of their work.

For this reason, it is essential that careful consideration be given by higher authority to what is permissible and what is not in a National Park, and that very distinct rules and guide-lines be laid down in this respect, which would assist smooth working relations between the Warden and the research workers of a National Park, considerably lessening the grounds for mis-understanding. In Tsavo, the Research Co-ordinating Com-mittee, representing the view of both factions, certainly went a long way towards achieving this need.

CHAPTER 16

Ntharakana

THERE are now many places on the steep slopes of the Yatta
Plateau in the northern area of the Park where crystal clear
water trickles from beneath the rugged lava boulders to create
a little oasis of lush green amongst the parched landscape. Ferns
and moss fringe the sides of these small streamlets, as the water
slowly trickles through the depressions made by the feet of
heavy animals. Wild creepers and flowers adorn the banks in
colourful profusion, while painted butterflies and scarlet dragon-
flies dance above the water like flamboyant fairies, and water
beetles dart in the miniature pools like small dark torpedoes.
Birds nest there, singing and chattering as they weave their
intricate homes on the ends of swaying stems overhanging the
streams. Such places are filled with a lazy peace and exude an
atmosphere of tranquillity as the wild creatures come to drink
and rest in the cool surrounds after a long and thirsty trek.

It was at one such place, known by the Wakamba as
Ntharakana, that David decided to build a comfortable retreat
overlooking a spring that oozed from beneath a massive boulder,
gathered momentum and cascaded musically over a sheer rock
face to meander gently on down the slope in a series of shallow
pools, until it disappeared again into the earth near an enor-

195

mous baobab tree. He designed the Ntharakana Blind to blend in with its surroundings as much as possible, and built it on two levels; six small double cabins, a flush toilet and small bathroom downstairs and a kitchenette, dining-room and viewing balcony above. The entire building was faced in natural rock while cedar offcuts panelled the concrete ceilings, giving the rooms a warm and cosy atmosphere. Shutters that could be lowered exposed a breathtaking view of hundreds of miles of wild, remote and unspoilt country – the home of giants: giant elephants and giant baobabs, for nowhere else in the Park were these two so plentiful. Behind the building, huge boulders rested in rugged splendour on the steep slopes of the Yatta, as though hurled by some herculean force in more turbulent times to remain as silent sentinels to a scene enacted since the beginning of time. From the cabins, one had only to raise one's head from the pillow in order to get an unobstructed view of the spring below, while an artificial moon situated on the balcony bathed the pool in a pale yellow light on dark nights. Most conveniently, another little spring higher up the escarpment provided clear water which was piped by gravity to a storage tank on the top of the Blind roof, and ensured not only a continuous supply for domestic purposes, but a cooling, gurgling sound, as it trickled from the overflow down the slope.

The Blind was really intended as a facility for those more adventurous tourists, who sought an exclusive camping safari in the northern Wilderness area, which was closed to the general public. But, it also afforded a haven, when not in use, for weary Wardens, who could occasionally steal away to soak up the solitude and silent peace of the place. For a Warden's life is not always as glamorous and idyllic as it is usually portrayed, at least, not in the Kenya Parks, where he is expected to be not only a naturalist, but also a master builder, roadmaker, designer, engineer, mechanic, pilot, policeman, public relations officer, overseer, accountant, administrator and Game Warden. His duties are as exacting as they are diverse, and involve the shouldering of enormous responsibility. They are often fraught with disappointments, irritations and frustrations from which there is no escape, for a Warden is expected to be on call twenty-four hours a day and seven days a week. His day begins at 7 a.m. with the daily radio call when all out-stations and patrols report to headquarters and receive their orders. Meanwhile, a long queue grows longer every minute as more and more people line up with their problems. The following, selected from topics requiring attention before breakfast one morning, will illustrate the start of a Warden's day.

'Excuse me, Sir – 10 lbs of nails are missing from the Store. They must have been stolen by someone.'

'My wife has gone mad and is running around in the nude. I would like twenty days' leave to take her home.'

'The blades won't fit the plane, Sir.'

'The tractor has broken down.'

'Manyani Gate is out of water, and there is no lorry to take more out there.'

'The painters have finished their work. What shall they do now?'

'A buffalo is chasing the visitors at Aruba Lodge.'

'The petty cash won't balance, Sir.'

'The borehole pump has broken down, and the pool is nearly dry.'

'They've sent the wrong spares for the grader.'

By this time, it was 8 a.m. and time for the radio call to Nairobi headquarters and other Parks.

'Your Quarterly Report is now six months overdue – we must have it by Monday.'

'Mr So and So will be visiting the Park. Please give him all the help he needs.'

'Your estimates for the coming financial year . . .'

Following this, David usually attempts a getaway from the office, but is pursued by a long 'tail' like the Pied Piper, with people trying to present their problems as he strides around the workshop allocating work. By nine o'clock, his responses are beginning to betray signs of exasperation, until finally, in des-peration, he turns on the 'tail' like a bayed up animal, and snarls.

This has the desired effect and the 'tail' begins to slowly dwindle, leaving only the boldest and most persistent to brave another attempt, until they too take the hint and drift off. By this time, the odd tourist has started to make an appearance with the 'Where are the lions?' type of queries, and then some of the scientists need placating. 'The road to my house is very rough. Could you do something about it?' and 'Could you do some flying this afternoon? I must complete my monthly count.' or 'Could I have a Ranger to accompany me?'

Breakfast brings a brief respite before the day's work begins in earnest, and I have learnt that it is also a very bad moment to present *my* problems!

To be able to retreat occasionally to Ntharakana, knowing that there was no possibility of anyone intruding, helped to keep one sane. There David could truly relax in a wild place, at one with the creatures that also came there. Time stood still, and the air was as though it had been breathed straight from Heaven itself. It was thrilling to sit on the rocks above the favourite watering places and quietly watch the elephants drink and bathe, unaware of our presence although we were so close that we could almost touch them. One became completely absorbed by these magnificent animals. No creatures are so peace loving and tolerant, so gentle and dignified, so sociable and sensible as elephants. They will help each other when maimed or injured, protect each other in a tight corner, look after each other if need be, and show consideration towards one another, living together in peace and harmony. Herds will meet, greet each other with dignity and move on. Strange cows will adopt and mother a needy orphan, accepting it and tending it with as much love and care as they would their own. Even strange bulls will take a hand in ensuring the safety of a small calf, as was demonstrated by Samson when he retrieved an orphan and brought it home. We have also seen a bull, which was in no way connected with the herd, place his tusks beneath the stomach of a small calf that had left its unit to venture deeper than usual into a waterhole. In fact, this baby was enjoying itself immensely, but until it was safely reunited with its mother, the bull, with gentle concern, stood by, prepared to support it should it get into difficulties. Large bulls will often allow a calf to share a drinking hole in a riverbed, although they will repulse any adult expecting this liberty. Elephants, although unmatched for strength, are never arrogant, and will extend great tolerance and consideration to other species, even preferring to stand aside rather than risk a confrontation with a cantankerous rhino or an old bull buffalo. David once witnessed a ridiculous scene from the air, when five elephants were grouped around a tree in a semicircle with only their heads in the shade and their bodies exposed to the burning midday sun. A closer look revealed the reason for this rather strange sight. A sounder of warthog were stretched out luxuriously enjoying the deep shade at the base of the tree, and the elephant were prepared to accept this situation, although it must have caused them a good deal of discomfort.

Although young bulls frequently indulge in playful bouts of sparring, only very, very seldom will elephants fight seriously, and then usually only if the two contestants are evenly matched in tusk size. More often than not, they seem to know their place instinctively, and a slightly smaller bull will step aside to allow the larger undisputed right of way. But, when a real fight does occur, it must surely be unmatched in the animal kingdom for power and sheer ferocity, and will often result in the death of one of the parties.

How can we gauge the depth of their intelligence? What goes on in that massive head, for instance, when coming across the carcase of a fallen comrade, they will deliberately pull the tusks from the sockets and will carry them away and sometimes even smash them to pieces? And what, I wonder, was going through the minds of a herd of elephant in Uganda, who stood outside the store that housed the feet of hundreds of their cropped companions, and shovelled earth into it through a narrow opening until the feet were partially covered with soil? What was the motive behind the two elephants in Tanzania, who are reported to have supported and guided an aged blind buffalo for some two days? Why, I wonder, do elephant often cover bodies with bush and grass, bodies not only of other elephant, but of other animals as well, and even humans? These questions must remain unanswered, but knowing all this, who can then say that animals are not capable of the same emotions and feelings as we ourselves? And elephant are by no means unique in this, for buffalo too form lasting and loyal friendships, particularly in old age, and it is not uncommon to find an old bull accompanied by an inseparable companion, who guards him, protects him and will stay by him. On two occasions, whilst flying, David has seen buffalo actually standing beside the bleached bones of one long since dead. The first time, he dismissed it as coincidental, but the second time, diving low over the spot, he was surprised to find that the buffalo refused to leave the bones, tossing its head and pawing at the ground, apparently prepared to defend even the last resting place of a colleague. Perhaps the lingering memory of a special relationship drew this buffalo back again to the only tangible thing that remained of his friend, in just the same way as we humans return to the graveside of a loved one. Incidents such as this certainly provide food for thought, and I wonder how many people, when shooting for pleasure, spare a thought for the sorrow and misery inflicted on those left behind as the result of their actions? For every animal that dies, there are others that care very deeply. The little dikdik who has suddenly lost her mate, or an old bull buffalo who will mourn the loss of a loyal friend. If only more people really understood animals, fewer would feel any desire to kill them. Reflections of this sort, in a setting like Ntharakana, made one feel very aware of the reasons for the isolation of man from the rest of the animal kingdom, for it is not difficult to understand why it is that he is hated by all other forms of life.

On the way back from Ntharakana, one day, we stopped for a while at Kiasa waterhole to watch some twenty-five elephant bathing and romping in the rain-filled pool. As we watched, one of the bathers suddenly gave vent to a terrified scream, and literally exploded from the pool in a sheet of spray, while the others, not knowing what was going on, also took to their heels and trumpeted in sympathy. The cause of the disturbance then popped up in the middle of the pool and gazed solemnly around: a solitary hippo, who must surely have felt a sense of achievement at having succeeded in routing the boisterous intruders who had disturbed his peace. Obviously, one of the elephant had unwittingly trodden on him, and one can appreciate the shock it must have been to feel that slippery shape heave protestingly, for Kiasa waterhole is the last place one would expect to find a hippo, being dry for three-quarters of the year. This hippo must have trekked from the Galana River, a distance of some twenty-five miles, or from the Athi River, which, although closer as the crow flies, would have entailed climbing over the Yatta Plateau. And how did he know he would find water at the other end, anyway?

During such absences, our orphans were very ably cared for by Mavis and Philip Hucks, an elderly couple who had retired from farming and whose love of wild animals and passion for wild places had tempted them to volunteer to undertake a botanical collection of the plants and trees of Tsavo in exchange for caravan room in the camping ground. Thanks to their valuable contribution to Tsavo East, the scientists now enjoyed not only a working Herbarium, where a pressed, named specimen of every plant collected was filed under the appropriate family, but also a unique collection of over 2,000 colour slides of the living plant itself for reference purposes, with details of fruits, flowers, bark, etc., representing what must be one of the most comprehensive and unique botanical collections ever made in any one area. Several species new to science had been discovered. The plants of Tsavo were as varied and as interesting as the animals themselves, for some species only flowered once in every five or six years, some for only ten minutes at dusk, but with infinite patience, Philip and Mavis laboriously recorded most of them for posterity. They lived in a caravan built by Philip himself on the chassis of a Bedford lorry. It was beautifully equipped and completely self-contained, and like Doctor Dolittle, the Hucks had a special way with animals. Birds of every hue, shape and size hopped around the caravan in search of the nuts and crusts of bread which were there at all times for their benefit. Some bolder residents would even take a morsel from an outstretched hand, while others like the hornbills would come and demand their share by knocking with their beaks at the window of the caravan. Tree squirrels, ground squirrels, genet cats and mongooses; all were regular callers and were fed by the Hucks. The waterbuck, who sought the safety of the open camping ground during the hours of darkness, found added protection near the caravan, and would graze peacefully and unafraid, seeming to sense the sympathy these particular humans had for them. A pale little treefrog settled down on the windowsill to see the dry season through, while a gaudy, striped spider, who lived in the centre of a gossamer-like web anchored to the caravan and a nearby tree, was made equally welcome. A little male dikdik, who seemed to be somewhat of an outcast from the others, sought comfort and shelter in Mavis' flowerbed; waterhog came to wallow in the small pool nearby and even a wild rhino made use of this facility each night. Other regular visitors were our own orphans, who, on their way down to the river, were certain of a warm welcome, a drink, something special to eat, and a friendly pat, making a stop at Chez Hucks part of their daily routine.

It was therefore not surprising that Bill Travers and Virginia McKenna, having seen the Hucks' abode only once, chose it as the perfect setting for their film *An Elephant called Slowly*, the story of which centred round a little elephant who befriends Bill and Ginny while they are caretaking a friend's bungalow in Africa. Unfortunately, on this occasion, we were unable to provide a suitable animal star, for Eleanor and Kadenge were both the wrong size and were too large for the part. Another little elephant which had been trapped by the Game Department and was destined for the London Zoo as a replacement for the elephant called Dixie, was therefore hired for the film and sent down to Voi. She arrived, tense and bewildered, with no reason for either liking or trusting human beings and with a determination to charge anyone who ventured within range. Although she was only very small by elephant standards she was nevertheless quite capable of crushing a person against the bars of the stockade, being possessed of the enormous strength of her kind, and an abundance of courage as well. As she was, she was completely unmanageable and of no use to the film until she had learnt not only to respect her captors, but to love and trust them as well.

The task of befriending Pole Pole, as she was called, meaning 'Slowly' in Swahili, fell to David, who had plenty of experience in this field, and after only two or three days of kindness coupled with firmness, it was possible to enter the stockade and fondle the little elephant in complete safety. Within four days the gate to her enclosure was opened, and she was allowed out, following Bill and Ginny because of a desire to be near them, having learnt to regard humans as friends rather than enemies.

Work on the film could now proceed, but one scene in particular was of special interest to us; the introduction of Pole Pole to Eleanor and Kadenge, who in the story appear as the friends she brings with her to the Travers' abode.

Eleanor, with her extremely strong maternal instinct and the gentle characteristics of her kind, would, we knew, be enchanted with Pole Pole and want to adopt her on the spot. We knew too that Kadenge would welcome the newcomer, although perhaps with more exuberance, and Rufus was quite accustomed to consorting with all sorts of creatures.

The cameras were positioned and the orphans herded nearby. Pole Pole was quick to sense their presence, and trumpeted several times, which brought Eleanor and Kadenge at the run. Eleanor, overjoyed and beside herself with excitement, greeted the little stranger with great warmth, rumbling lovingly and fondling Pole Pole protectively with her trunk, while Kadenge rushed around trumpeting with his ears spread out and his head held high, obviously also delighted at this unexpected turn of events. Rufus, so absorbed by the thought of sweets, didn't appear to even notice the addition of another elephant to the orphan herd. Thereafter, Pole Pole enjoyed the company of the orphans after she had completed her day's work, and became transformed, no longer cut off from her kind in hostile and strange surroundings, but an animal who had learned to live again, happy, playful, confident and unafraid. From Eleanor's large heart poured all the love, protection and attention that Pole Pole needed, and so, for two or three brief weeks, little Pole Pole was completely happy.

But, as the film drew to its conclusion, a feeling of despondency fell over everyone with the knowledge that Pole Pole would have to endure another painful parting, and having tasted freedom, become accustomed to being confined again in a distant land. The idea haunted Bill and Ginny, and nothing would have given them more pleasure than an opportunity to purchase Pole Pole's freedom, so that she could remain in Tsavo with Eleanor and be spared a Zoo existence, but alas, the authorities would not agree, for Pole Pole had been promised to the London Zoo, and was booked to fly to London.

We did what we could to make her last day in Tsavo particularly enjoyable. She lounged in the luxury of a mud wallow, soaking up the sun and plastering the red earth all over her body; she roamed the riverine forest, selecting delicacies for herself that she would probably never taste again; she dozed peacefully under a huge spreading thorn tree in the heat of the day, pressed close to Eleanor, and then a gentle shower fell in the afternoon to place the finishing touch to a last perfect day, scenting the air with the fragrance of dampened earth and making the leaves glisten silver in the soft afternoon light. Pole Pole frolicked and rolled and played with all the customary sense of elation that accompanies the first falls of rain in a thirsty land, and finally, as the sun sank below the horizon in a fiery ball, and the sky took on the delicate tints of a traditional African sunset, all the orphans filed slowly home. Watching them enter their stockades, we couldn't help being enveloped by an aching sadness for little Pole Pole, whose natural existence was to be sacrificed for the pleasure of people.

It is times like this, when an animal rises above being just another animal, and takes on the status of an individual, that one can be accused of being sentimental with justification. No good conservationist need ever be ashamed of this trait. It simply denotes a sympathy for animals, and no one working with animals should lack this essential quality.

It might seem ironical, and perhaps even ridiculous, to try to rear a baby elephant, for instance, when an overpopulation of that species is our most pressing problem, and it would certainly be stupid to embark on a deliberate campaign with that object in mind. But, when providence confronts one with a baby animal in dire straits, it would be a callous individual- indeed, who could turn away and leave it to its fate, for im, mediately it becomes an individual, and as such must be con- sidered in a very different light. Everyone agrees that the world is overcrowded, yet who would volunteer to be the first to rectify the situation! Under such circumstances, even the most logical arguments become invalid.

The following day, once more bewildered and tense, Pole Pole was driven away in the crate in which she had come, and several days later a picture in the local paper of her being taken aboard the aircraft destined for London brought it home even more forcibly that she was now well on her way.

Eleanor missed her very keenly, particularly as her departure was closely followed by that of Kadenge, who had to be per- suaded to join his own kind. She took to joining the wild herds herself, for periods of up to three weeks at a time, to reappear again suddenly, as placid and as gentle and affectionate as ever, resuming the daily routine with the other orphans just as though this had never been interrupted. There were many times that we thought she had left us altogether when the days stretched into weeks, but she was always drawn back in the end, torn between her natural instincts and her affection for us.

Her loneliness was eased somewhat, however, and her motherly instincts appeased, with the arrival of another orphaned elephant; this time a bull, whose mother had paid the extreme penalty for trespass on the neighbouring Sisal Estate. He was immediately handed over to Eleanor, who again adopted this calf without hesitation, lavishing on it all her customary love and attention, and instilling in it sufficient confidence for it to realize that the humans with whom she fraternized so freely posed no threat. We called this elephant 'Ndara', after the place from which he came. He was in good condition on arrival and required no special attention on our part, Eleanor doing all that was necessary. He kept close beside her at all times, being herded with the other orphans and fitting into his new life with perfect trust in his new 'mother'. But Eleanor still felt the call of her kind, and she again took to periodically joining up with the wild elephants, although for shorter periods than before, taking Ndara with her and bringing him back again when she returned. We were surprised that he didn't take the opportunity to switch loyalties and attach himself rather to an adult cow, who, one would imagine, would be a more attractive fostermother, but he was devoted to Eleanor and stuck to her through thick and thin, although she was only twelve years old and not yet fully mature.

These periodic excursions were shortlived though, and ended abruptly one day with the arrival of yet another baby elephant, whose mother had been shot by a party of unscrupulous Asian hunters just outside the Park boundary near Sala. This unfortunate little elephant, who was barely old enough to browse for himself, had remained beside his mother's inert body for three days before being rescued by the Game Department and brought to us in Voi. The arrival of the Game Department was timely indeed, for had they delayed a day longer, it is likely that a smaller corpse would have joined its mother's remains. As it was, the calf was extremely emaciated and weak on arrival.

As the little elephant was being unloaded, Eleanor, who had been put in her stockade for the night, was straining at the bars, reaching out her trunk to its fullest extent in an attempt to touch the calf, rumbling reassurance as it was being hustled along. Opening the gate to her enclosure, we simply pushed the calf inside, and it was stirring to witness yet again the boundless compassion of elephants, as Eleanor gently drew the baby to her with her trunk, fussing over it and coaxing it beneath her tummy, rumbling softly to it all the time, accepting it without reservation as though it was, indeed, her very own.

What other species on earth has such depth of feeling and consideration for its own kind? I'm sure there is none that can begin to compare with the elephant.

The little elephant responded immediately, and glued itself to Eleanor, drawing obvious comfort from her as it huddled between her forelegs. It was at this stage that we left, knowing that the newcomer was indeed in good hands.

It had been difficult to judge the approximate age of the calf in the dark and we fervently hoped that it was old enough to browse, for very young elephants under three feet in height, are practically impossible to rear, being extremely delicate and unable to tolerate anything other than elephant's milk. Calves who have begun to browse, even a little, seem able to tolerate milk better, but even so, it had been our experience that milk to a baby elephant should be avoided at all costs. Our new elephant was a borderline case, but we decided that with special attention he might just be old enough to make the grade, although he was not very adept at using his trunk yet, and would need some assistance in this respect. The orphan attendant was therefore instructed to hand feed the little elephant as much as possible, gathering foliage for it until it became more proficient itself.

The real name of the little elephant was Kulalu, taken from the place near which it had been found, but it earned another name bequeathed on it by the attendants: 'Bukaneza' meaning 'the weak one' in the Boran dialect. Although all went well for two weeks, and we were confident that Bukaneza would survive, he suddenly collapsed half way up the hill at the back of the house, and the attendant hurried down to report that he was dead. We questioned the bearer of these ill-tidings further, and extracted the information that the elephant was actually not yet quite dead when he left, but was sure to be so now, because it was unable to get to its feet.

Obviously there was no time to lose, so we hurriedly mixed a strong solution of glucose, and hastened to the place where Bukaneza was lying as though indeed dead, while Eleanor milled around in extreme agitation. Only a very slight heaving of the little elephant's flank as he breathed spasmodically indicated that all hope was not lost, and that the flame of life still flickered feebly in his small frame. David and the Ranger struggled to raise him to his feet, but his legs kept crumpling beneath him and he seemed beyond being able to help himself in any way at all. Meanwhile, Eleanor, who must have understood what they were trying to do, knelt down, and placing her trunk and tusks beneath Bukaneza's stomach, lifted him to his feet with the greatest of ease, and supported him in a standing position with her trunk as we offered him the bottle of glucose. To our joy he downed the contents avidly, seizing onto the large rubber teat like a veteran bottle feeder, and sucking strongly. This gave him a little more strength in a very short time, but we realized that the improvement would only be temporary unless we could get some concentrated nourishment down him as soon as possible. It was at this stage that we decided to give him milk in the form of a calf supplement with a low fat content sold under the name of 'Trilk'.

We thought we would never fill Bukaneza up. He gulped down bottle after bottle until the entire gallon I had mixed had disappeared, and every time we lifted him Eleanor did her bit by helping to support him as he fed.

The lorry had meanwhile been despatched to Voi township to search for oranges and sweet potato tops. We thought the appearance of the oranges would be the moment when Eleanor's liking for them would outweigh her concern for her sick foster-child, for normally, whenever any oranges are produced, she barges and shoves everyone aside in her eagerness to get at them and stuff as many as possible into her mouth in the shortest time. It was surprising, therefore, that on this occasion, she stood by quietly and watched them being cut and fed to Bukaneza, demonstrating amazing self-sacrifice, and intelligence, for it was as though she recognized that Bukaneza's need was the greater on this occasion. Such unselfishness could not go unrewarded, so David saved some of the largest and juiciest oranges especially for her.

After allowing Bukaneza to rest until the early afternoon,

Elephant on the Galana

we decided that an attempt must be made to get him down the hill and back to the stockade. With Eleanor's help we achieved this, although it took the rest of the afternoon to coax him very slowly down, with long rests in between.

Thanks to Eleanor's care and encouragement, Bukaneza recovered completely. The extra care and attention he needed when he was so very weak seemed to forge an even stronger bond between them, and although Eleanor was to acquire many foster-children in the future, Bukaneza remained her firm favourite and was always accorded preferential treatment. A gallon of Trilk, served in an old white plastic jerry-can, the opening of which was very conveniently just right for the monstrous veterinary teat, supplemented his diet and helped to tip the balance. Soon he lost his gaunt, hollow look and took on the shape of a small barrel. He also began to demonstrate an impish character packed with personality. With amazing accuracy he could calculate feed times, and would wait tensely until the Ranger carrying the precious bottle appeared in view. He would then charge down the hill from the stockade to intercept the Ranger, who had to nimbly dodge aside at the last minute, and run as fast as he could up the hill with Bukaneza squealing frantically close on his heels. If the bottle was delayed, Bukaneza would work himself into a veritable frenzy, rushing up and down the hill trumpeting like a small steam engine, until it appeared, and come what may, rain or shine, he was always back home for his bottle, and Eleanor, who would never leave him, sacrificed her desire for freedom, and had to come too.

The thing that Eleanor, and indeed all our elephants, liked most of all once they had acquired a taste for it, was lucerne, and sometimes when conditions in the Park were very dry and food hard to come by, we would order a truck load from a farm upcountry in order to supplement their diet at night. Bukaneza, who had always shared Eleanor's stockade, never got a look in when lucerne was in the offing, for Eleanor would jealously guard the pile, munching her way greedily through the lot without so much as allowing poor Bukaneza even a mouthful. If he tried to snatch some, he stood to get severely reprimanded and shoved aside. Finally, we decided that he should have his own stockade where he could enjoy his share of the lucerne in peace, so he was transferred to the vacant enclosure next door to Ndara. However, this turned out to be not a wise move, for although all was well until the novelty of having his own lucerne wore off, he then felt the desire to be reunited with Eleanor for the night, and set about trying to squeeze himself between the horizontal bars. He succeeded only in getting his head firmly wedged. Chaos followed, and while Bukaneza yelled and the herd boy frantically tried to free him, Ndara attempted to crawl underneath his horizontal bars, and Eleanor set about trying to climb over the top of hers to come to the rescue of her baby. The noise was indescribable, and all three elephant were frantic, until somehow Bukaneza's head got pushed back in, and he was allowed to join Eleanor. Even so, the upheaval upset them all so much that the next morning they appeared physical wrecks – hollow and hungry, having lost their appetites for the rest of the night.

Thereafter Bukaneza enjoyed the best of both worlds, with his lucerne to himself in his own enclosure, but with the door left ajar so that he could join Eleanor whenever he wished. Fortunately, the bars surrounding her stockade were sufficiently wide apart for him to be able to crawl through without getting stuck.

Today, Bukaneza is a cheeky little character, who, confident of his privileged place with Eleanor, can afford to throw his weight around with the other orphans. Occasionally Ndara becomes exasperated and can't resist taking a dig at him, but very often falls foul of Eleanor in doing so. We have seen Eleanor, in response to Bukaneza's protests, become extremely angry with Ndara at such times, and chase him around trying to tusk him for bullying her baby.

A demanding character too, is Bukaneza! When he gets a thorn in his foot, he lifts his foot up and expects the attendant to remove it; and when it is bottle time he expects it to be produced – on the dot!

CHAPTER 17

Tourist Facilities

THE tourist industry since Independence was rapidly becoming an extremely important facet of the country's economy, rivalling and finally actually overtaking coffee and beef as the principal foreign exchange earner.

It is common these days to hear how the former Colonial Government neglected wildlife. But, in all fairness, who created the National Parks in the first place, and began the development of them, carrying them for many years through a period when the returns from tourism were negligible and they represented no more than a dead loss? The former Government surely deserves some credit for this, and some gratitude too, for, unaided, it was caretaker to a legacy, the benefits of which are now beginning to be realized.

Lodges had started to spring up within the Parks since Independence to accommodate the flood of tourists who began to pour into the country. Kilaguni Lodge, in Tsavo West, pioneered the more sophisticated type of Lodge where all the amenities, including swimming pools and full catering facilities, were provided. This proved so popular that another luxury Lodge was constructed, sited again in Tsavo West on Ngulia Mountain, and then finally, attention was turned to Tsavo East.

It has been found in many other countries, that development of this nature, which has a tendency to 'snowball' and bring with it many aspects that are not really desirable in a National Park, such as staff villages, disturbance, litter, etc., should be confined to the periphery of a Park, and not be established actually within it. There have been instances where the facilities erected for the comfort of tourists at a place of beauty have ended up by destroying it, and for this reason, had he been allowed a say, David would have preferred the new Lodge to be sited outside the Park Gates. On the other hand, it could also be argued that an investment of this nature within the Park was a form of insurance, and made the Park that much more secure.

There is a hill behind the one on which the Park headquarters lies. It rises steeply on the eastern side in a series of giant granite boulders and enormous slabs of solid rock. The western approach is gradual from a tongue of higher ground that skirts the back of the hill. This hill is known as Worsessa, which means 'rhino' in Waliangulu, and it has always been one of our favourite look-out points, for it commands an uninterrupted vista of a vast stretch of flat country, with the deep blue massifs of Ndara and Sagalla to the right, and the thin ridge of the long Yatta Plateau on the far horizon. Sunlight plays tricks with the colours on the plain in the dry season. Dark blue-black shadows cast by passing clouds crawl slowly across the pale yellow grass, areas of impeded drainage take on a slighter lusher hue and are tinged with the faintest suggestion of green. Tall melia trees fringe a distant waterhole while other dormant types take on fascinatingly twisted shapes that provide endless variation and have a stark attraction of their own. Patches of rich terracotta soil offer blended contrast in the form of stately anthills, and all this gradually loses its identity in the distance, as it merges in a confusion of blues and mauves, all with a slightly smokey effect, until the pure blue of the sky stoops to join the line of the horizon.

The boulders on Worsessa Hill, too, are breathtakingly beautiful, and growing amongst them is an assortment of small shrubs, some of which are adorned with blossoms as unique and varied as the surroundings in which they stand.

It is a place of quiet tranquillity, not unlike Ntharakana, and many were the times when we sat on the rocks at the top and gazed lost in thought at the scene below. A scrutiny through binoculars never failed to unearth rhino, buffalo, elephant and many other smaller creatures feeding peacefully on the plain. Rock rabbits and mongooses lived in crevices amongst the boulders, and once a lion, angry at the interruption, jumped from below an overhanging rock nearby with a low, spine-chilling growl.

Not without some regrets, therefore, we decided that Worsessa should be the site for the new Voi Safari Lodge, realizing that yet another peaceful spot would disappear forever. Sitting there, pondering this, David remarked,

'I expect the day will come when I will walk into the Lodge, and the manager will look me up and down critically, and say that gentlemen must dress for dinner!'

The construction work of the Lodge took over a year, but the result was a magnificent piece of architecture designed to fit the site, with advantage being taken of the natural boulders, incorporating as many as possible into the building. So skilfully suited to its natural setting, faced with the red rocks of Tsavo, it blended into the hillside completely. It comprised fifty double bedrooms, all overlooking an artificial waterhole, a swimming pool and all the amenities of a modern luxury hotel. The landscaping of the grounds was also a work of art. Little pools and waterfalls brought coolness to the surroundings, and every plant, shrub or succulent in the gardens was indigenous to the Park itself. Brilliant desert roses ranging from deep crimson to pale pink stood out to give the setting colour, other small succulents had flowers that opened to reveal a pattern resembling a Persian carpet, attractive grasses and giant aloes graced the slopes, while wild creepers, flowering shrubs and trees were skilfully positioned to show them to maximum advantage. The result was something of which everyone in Tsavo East could well be proud.

But, it took the game quite a long time to become accustomed to the presence of the Lodge, and to gain sufficient confidence to venture to the pool. Gradually, the more adventurous led the way, and others then followed, until thin game trails became established, converging on the pool from all angles and according it recognized and seasoned waterhole status. Thereafter, elephant filed in throughout the dry season, a resident herd of some 500 buffalo watered there regularly, pushing and shoving each other aside for the chance of a drink, and rhino, too, became regular callers. Lion also made use of it, ambushing and killing their prey in full view of the spectators at the Lodge on more than one occasion. Warthog, baboon and waterbuck provided permanent interest at all times, and then the shyer, more timid creatures began to venture in; the elusive and graceful lesser kudu, eland, impala, zebra, giraffe, leopard and even cheetah.

There were other experiences too that were not so popular with the guests, provided by the lesser forms of wildlife; the incredible variety of insects of all shapes and sizes, and the occasional appearance of a cobra or puff-adder in the Lodge rooms, which never failed to cause quite a stir. One night a young lady guest burst from her room, stark naked, and fled screaming down the corridor, pursued (she thought) by a large dung beetle the size of a matchbox, which happened to be zooming along in her direction with a loud and ominous hum. It must have appeared a terrifying apparition, but was, of course, really quite harmless. Another time the manager looked up to see a cobra beneath one of the tables in a packed dining-room, and to his horror, four elderly guests about to take their seats around it. With commendable presence of mind, he leapt forward and snatched the chair away from the nearest old lady, just as she was going to plant herself on it. I often wonder what went through her mind in the split second before the manager was able to explain the reason! A good deal of confusion ensued, with waiters rushing around looking for missiles with which to tackle the unfortunate snake, and all the guests surging towards the furthest end of the room while it was being despatched, and the manager imploring them to keep calm!

Soon after the completion of the Lodge, tragedy struck one night, with the sudden death of Philip Hucks, who with Mavis lived in a caravan in the camping ground near the main Gate. As might be expected, his last thoughts centred on the well-being of an animal; a little wild genet cat, who came regularly to the caravan at night for food. 'Have you put Jenny's food out?' he asked Mavis, and then, the next moment, he had gone.

Philip was buried in the Park, to rest forever in the place he loved amongst the creatures he loved. And during the short open air service, a hornbill called in the tree above, and a herd of elephant browsed peacefully within view. Our sorrow at the death of Philip was too profound to be able to be expressed in words.

Philip was a real character, one of the brand of old-time Englishmen of the calibre that made Britain great. His experiences in the early days rivalled the most exciting of thrillers. He had been severely mauled by a leopard while hunting in Uganda, had been carried, semi-conscious and in dire agony, on an improvised stretcher by his faithful gunbearer and servants for several days before reaching the banks of the Nile. Here he was placed in a canoe, which promptly overturned in midstream, nearly drowning him in the process. Finally he was deposited on the opposite bank, more dead than alive, but managed somehow to drive himself to hospital in his car, which had been left at a trading post, and get himself to the carpark before slumping unconscious over the controls.

Weeks stretched into months, during which time he clung grimly to life with sheer determination. And all the time, Mavis, then his fiancée, who was made of the same stout stuff, puzzled over the fact that she had been without news for so long, and finally embarked on an expedition to Africa to determine the cause for herself. It took quite a long time to track Philip down, but she found him in the end in a remote Uganda hospital facing the possible amputation of both a leg and an arm as the result of the severe wounds he had received. But Mavis was not to be defeated. She set about getting him back to England, which in those days was not as quick and easy as it is in present times, and after many months of hospitalization, he slowly recovered and kept his leg and arm, although the leg was left permanently stiff and the right hand distorted. Added to all these handicaps was the fact that he had a weak heart, but what his heart lacked in physical strength, was more than ever recompensed by sheer guts, kindness, compassion and ability. Philip was extremely versatile in spite of all his disabilities. He would climb trees to pluck a flower, trudge through the bush for hours, carry out major repairs to anything needing attention, ranging from his Bentley to a lawn sprinkler, a camera to a cupboard. He was also a qualified draughtsman, a successful farmer, an excellent photographer; a fiery foe and an extremely good friend. And he was one of the few people who can claim to have run over a full-grown rhinoceros.

On this occasion, he and Mavis were on their way to the Sala area of the Park, where an unseasonal shower of rain had brought the bush into leaf and many of the shrubs into flower. Mavis went ahead in an old Land-Rover, which, judging by its appearance, must surely have been one of the first ever made. It had no windscreen or hood, no seats and very little comfort, but it was the Hucks' mode of transport for working purposes, the Bentley being reserved for the odd excursion back into civilization. Philip, following behind, was driving the old Bedford lorry which also served as their caravan.

As they were driving along through the Park, a rhino, which had been disturbed by Mavis passing in the Land-Rover, suddenly exploded onto the road behind her, and proceeded to give chase. She was completely unaware of this development, the rattling of the old Land-Rover drowning everything, including the rhino's snorts and puffs as it rapidly gained ground. Philip, on the other hand, saw it all, and fearing that at any moment the rhino would toss the Land-Rover and Mavis with it, he urged the old caravan forward at top speed in a desperate bid to divert the rhino's attention and avert disaster. The sudden deafening roar behind the rhino had the desired effect, and it spun round to face what must have appeared as a very formidable and determined aggressor. Although Philip hastily jammed on the brakes, sheer momentum carried the caravan straight onto the rhino, who was braced solidly in the middle of the road with its head down to take the impact. The collision brought the caravan to an abrupt standstill, and also knocked the rhino unconscious, wedging it underneath the front bumper.

Throughout all this excitement, Mavis continued to drive along, oblivious of everything except the pleasant scenery and the indescribable din made by her machine as it bucked over the corrugations in the road.

Philip got out to assess the situation, and a quick appraisal left him with the firm belief that the rhino had been killed outright. He was therefore busy concentrating on the damage to the front of the caravan, when he happened to notice that the rhino appeared to be showing signs of coming back to life, so, with his habitual stiff-legged gait, he hurried back to the driver's seat, and reversed the vehicle, which by now was beginning to heave like a ship on a rough sea. However, this freed the rhino, who then proceeded to set about one wing of the lorry in no mean fashion, before trotting off into the bush, apparently suffering no severe ill-effects apart from a sore head, and a very bad temper!

Before setting out that morning, a rendezvous at Aruba Lodge had been arranged to mark the end of the first leg, and it was here that Mavis was awaiting Philip's arrival with mounting concern and impatience. When he did finally turn up, also not in the best of moods, she was incredulous to hear of all that had taken place – and so frankly were we, when a report came over the radio that the 'Bwana Mzee' had run over a rhinoceros!

In a corner of this huge Park, an enormous natural boulder with a simple plaque and inscription inset, marks the last resting place of Philip Hucks forever, and the work that he left on the plants of Tsavo, immortalized in a working Herbarium and over 2,000 colour reference photographs, serves as a salutary reminder of unstinted dedication undertaken by one who chose to spend the twilight of his life doing something really worthwhile for wildlife. It is fitting that he should remain in the surroundings he so loved, and those that knew him are content in the knowledge that he would not have wished otherwise, and rests in Tsavo in peace.

CHAPTER 18

Water is Life

As a result of the changes in the habitat brought about by
elephants and fire, there had also been some interesting changes
in the distribution of certain species within the Park. For
instance, in the early days, the Somali ostrich and Peters
gazelle were confined to the area north of the Galana River
and were not seen in the southern portion of the Park. Although one of the early hunters claims to have shot Peters
gazelle near Buchuma in the very early days long before the
Park came into being, certainly when David first came there
was no evidence of their presence anywhere south of the
Galana, and it was not until as late as 1958 that this species
began to extend its range. Since then they have spread and
are now quite common throughout Tsavo East. It is only very
recently that they have moved into the Voi River valley and
become established near the western boundary. The one which
pioneered this new range was a young orphaned female given
to my brother Peter, but it escaped from custody and made

friends with some gerenuk near Irima Hill on the Mudanda
Road. As far as we knew, the nearest wild ones were at Aruba,
some sixteen miles away, and we felt rather sorry for this rather
forlorn little antelope whenever we chanced to see it. One day,
to our great surprise, another female had mysteriously appeared
and joined it, and four years later the group consisted of four
females, but still lacked a male. Now, I am happy to say, one
has arrived, to complete the first herd of this species in this
area.

There is no doubt that the predators, particularly cheetah
and wild dog, who normally chase their prey for some distance,
play the important role in Nature of dispersing animals, resulting in the extension of home ranges and the prevention of too
much in-breeding within the ungulate herds, and this is very
likely how those Peters gazelle came to be at Irima. Many
antelope are equipped with a means of marking their territories, and this too no doubt plays a part in the location of
their own kind. The handful of Hunter's antelope, for example,
which were translocated to Tsavo East in 1964 from the Tana
River, and which were subsequently released on the Ndara
Plains, did not remain together as was hoped, but scattered
far and wide. For several months only the occasional one was
sighted in different places with different animals, until the onset
of the rains, when the whole lot seemed to vanish altogether.
For two years there was no record of them, and it was feared
that the exercise must have been a costly failure, until 1966,
when a herd of eight Hunter's antelope was suddenly spotted
from the air on the Dika Plains. Banking steeply, David could
see that this little group included two or three yearlings, which
must have been born in the Park. Today the herd of these
rare animals numbers ten, while several males have splintered
from the main group and are occasionally seen amongst other
species or alone. So, the exercise was not a failure after all,
and it is wonderful indeed that in the immensity of Tsavo and
in strange surroundings, those few animals, many of which were
still adolescent when released, managed to trace each other
and weld themselves into a herd.

Obviously the ideal way to translocate animals is to aim to capture an entire herd, so that the animals are familiar with each other and will therefore be more likely to remain together. Soon after the arrival of the Hunter's antelope, twenty Grevy's zebra were moved to Tsavo from the Northern Frontier, and again, although these animals were kept in holding pens for several weeks, when the time came to let them out, they too dispersed in all directions, and for several years were seen only occasionally in small groups of twos and threes, or individually, usually amongst herds of the common Burchell's zebra. However, today, twelve Grevy's zebra have joined up to form their own unit, and amongst this herd are several foals born in Tsavo, while the odd solitary stallion and splinter groups of two and three animals continue to be seen regularly. Unfortunately, one of our precious Grevy's zebra fell victim to a hunting party when it wandered outside the Park boundary into a neighbouring hunting block, and when the hunter in charge felt that he had made a discovery worthy of positive proof and record.

The Somali ostrich, today, occurs all over Tsavo East and can be described as fairly common, whereas, when the Park was first proclaimed in 1949, two years passed before David even saw one, and then they were to be found only north of the Galana River. Impala, although common along the Galana River, first appeared in the Voi River valley only some six years ago; yet today they are becoming widely distributed and continue to extend their range, all the time pioneering new areas.

Another interesting development of recent origin has been the dramatic arrival of oribi, a species hitherto unknown in the Tsavo Park, and unknown also to the Waliangulu we have spoken to, although they do occur in the coastal belt in small numbers. This species was first recorded in Tsavo East in 1969, when we happened to be motoring across country over the Dika Plains area; an area which, incidentally, was once covered in thick commiphora bush. Suddenly David jammed on the brakes and groped for the binoculars.

'I can't believe it!' he exclaimed. 'I honestly can't believe it!'

Bouncing along in the tall grass were five small fawn coloured antelope that were quite new to me, with the white underside of their fluffy cottontails prominent and a black gland at the base of each ear – Haggards oribi. Since that day, they have been seen periodically. And not far from the Park's eastern boundary there is a remnant group of stately sable antelope, while opposite, on the north bank of the Galana, the kongoni's first cousin, the beautiful plum-coloured topi survives in small numbers. It is our sincere hope that Tsavo East, which can already boast a greater variety of large mammals than any other Park in the world, will one day be able to include these two varieties on its impressive list of some forty-eight different species. And what is more, near the Sala Gate on the eastern boundary, there is reputed to be something even more impressive – a mermaid, or dugong, living in the Galana River. Although one normally associates this creature with the sea, it is, of course, a mammal, and quite capable also of frequenting fresh water. The witness who testified to its presence could not have had better credentials either, for it was none other than the Mammologist to the National Museum.

One lone Grant's gazelle ram has suddenly arrived near the western boundary to the Park, and whereas this particular type of Grant occurs in Tsavo West, this is the very first individual to put in an appearance on the eastern side of the Tsavo Park.

There is no doubt that the work of the elephant had been beneficial to the grazers, many of which were now to be seen in sizeable herds. The stage had been reached when the increase in numbers was becoming more noticeable with every year that passed, for the long, slow, uphill haul to reach the hundreds from a handful had been surmounted. This was particularly true of zebra and buffalo, which were formerly found in only small groups, but which now occurred in herds of a hundred and more. It was our ambition to fill those newly created plains until they echoed the past profusion of wild life, so that here in a corner of Tsavo, there would again be great herds: herds of stately oryx with their scimitar-like horns, of beautiful striped zebra cantering across the plains in their thousands, of noble eland, graceful gazelles, impala and kongoni, and somehow this aim did not now seem quite so ambitious as it had twenty years ago. Even the Tsavo lion now roamed the Park in large prides. As many as twenty had been seen together on a number of occasions, which, in itself, may not sound so remarkable to those used to places like the Serengeti and the Mara, but for Tsavo this was a talking point.

Another amazing development had been noticed amongst the lions; the presence of several magnificent maned specimens in the Park, that bear no resemblance to the rangey, long-legged, mean-looking inhabitants one normally associates with Tsavo.

I had always believed that the absence of manes in the low country lions was due to the hot climate, and that our type of lions had indeed become a race apart from those proud, indolent specimens of the Masai country. It was often very difficult to differentiate between a male and a female at first glance, for the best the Tsavo lions seemed to manage was a few sparse hairs that could hardly qualify as a mane. Why then were they suddenly beginning to sport magnificent manes? Was this correlated in some way to the changes in the environment; perhaps to the better living conditions provided by the more open terrain, or were these handsome individuals also immigrants to the Park?

All these developments were, indeed, very gratifying, but the limiting factor in this arid region still remained water. In spite of the appearance of the mysterious springs, sources of permanent water were few and far between in the south eastern portion of the Park, where the greatest concentrations of grazers were to be found, and while the build up in numbers had been spectacular, none would deny, when considering the extent of the Park, that it was still poorly served when it came to numbers of animals, with the exception, of course, of elephant and rhino, and still during the dry seasons, many hundreds of acres of good grazing lay unproductive, while the country around the permanent water had to carry the entire load and often became trampled as the result. There was definitely a need to spread this load, which would not only raise the biomass to ensure maximum utilization of the habitat, but would go a long way towards increasing the tourist potential of the Park and consolidate the position of wildlife; an urgent need in view of the introduction of new drugs to combat trypanosomiasis in domestic stock. And whatever one may think about the provision of artificial water in a National Park, one had to be realistic, for a number of ranching concessions had appeared along the periphery of the Park, and cognizance had to be taken of the water development being undertaken in these areas, making it important that the Park did not remain in isolation. If we hoped to keep our game within the Park, parallel water development had to be embarked upon, otherwise the animals would be tempted to stray out of the Park in search of better grazing, and would become decimated by poachers and licensed hunters alike. It was a disappointment, therefore, when a request for funds for this purpose from the Wildlife Society was turned down by the Scientific and Technical Committee on the grounds that further research was necessary in order to assess the desirability of increasing artificial water supplies within the Park, and the Aruba dam, which is sometimes subjected to severe trampling in times of drought, was quoted as an example of the harmful effects such a measure could have on the habitat. The fact was, that had there been no Aruba, there would have been an even greater concentration on the river and even more trampling there, whereas this problem would have been eased had there been a dozen Arubas and not just one!

It was the candid opinion of a great many people who were members of the Society that far too much of the funds, donated by people with a genuine desire to improve the lot of wild animals, was channelled instead into nebulous drawn-out research projects, whose benefits to the cause of conservation, management of Parks, or the animals themselves, were very difficult to see.

As far as water development in Tsavo East was concerned, there was, unfortunately, not the time to embark on a lengthy

Voi Safari Lodge

survey of this nature, for our neighbours were not bothering with this luxury and were busy installing watering points for cattle in a great many places. Quite apart from this, it would take many years for any sort of answer to emerge, and in the meantime a lot of game would move out of the Park. David therefore decided to resort to trying to raise the wherewithal piecemeal as best he could from private donations and from the Park's own meagre resources, and, as and when a small sum had accumulated, he set about trying to improve the distribution of permanent water which was something he considered vital to the long term survival of the Park.

The water development carried out to date ended with the Aruba dam, and four bore-holes, one on the Ndara Plain, one east of Aruba and two on the Tiva floodplain in the northern area. Bore-holes, on the whole, had been disappointing in Tsavo East, which lay mainly on the basement complex and where chances of striking water were confined to the weathered zone overlying the basement, or from the fractures or jointing in the rocks themselves. Several attempts in the past undertaken in the basement complex had proved fruitless, but there had been one notable exception that had confounded all the experts; the Ndara bore-hole, which yielded 5,000 gallons of water per hour from a depth of only 125 feet.

Bore-holes definitely possessed certain advantages over surface supplies in that they could be controlled and were not dependent upon the rainfall. They were also a very important substitute for surface water in times of drought, it being feasible to keep some of the natural pans topped up with piped water, ensuring that animals were not stranded and relieving the pressure on the rivers. Because drilling costs were high however, and the progress slow, it was planned to explore as wide an area as possible by means of a small Drifter Drill capable of probing a $4\frac{1}{2}$ inch diameter hole to a depth of 250 feet in only 8 hours. In this way, it would be possible to plot the most promising points and be more certain of success before embarking on an expensive bore-hole.

But the main objective was to concentrate on the shallow

natural pans made by elephants, which were to be found throughout the Park, but which seldom exceeded a depth of 4 feet and dried up very quickly. Assuming that these pans were replenished twice a year during periods of good rain, they could be made permanent if deepened to a depth of 20 feet, the loss from seepage, evaporation and utilization by game amounting to approximately an inch a day. The target was to excavate over fifty such places at strategic points; a major undertaking with the limited resources at our disposal, but work was put in hand on the first one, sited on the plains towards the southern boundary with a view to relieving the load on the Ndara bore-hole and Aruba dam in the dry season.

An incredible quantity of earth seems to collect from such an exercise, and disposing of all this soil without resorting to unsightly dumps that detract from the natural scene, always presents a problem. In this case, the difficulty was overcome by raising the level of the road bordering the waterhole by several feet above the surrounding country, serving a twofold purpose; better drainage from the road itself and a better vantage point from which to view game at the pool. The road also was aligned with an eye to the drainage lines in order to channel as much run-off as possible into the waterhole.

Another idea was to install a series of small concrete or gabion weirs across some of the luggas in order to raise the water table and conserve as much water as possible in subsurface dams, which could be exposed by elephants and made available to other creatures when needed.

All this was only part of a comprehensive water development plan, but the total cost of this pipedream was something in the region of £180,000; a figure likely to daunt even the boldest of donors. It was nevertheless a goal, albeit rather an ambitious one, for the first waterhole took almost a year to complete with the rather antiquated machines at our disposal. However, fortune favoured Tsavo East one day with a substantial donation in the form of earth-moving equipment made unexpectedly by a wealthy American, who was so moved by the sight of a lone bull elephant standing dejectedly at a caked water hole, that he resolved on the spot to do something positive to help. Thanks to the compassion of this man, wild animals now enjoy four additional permanent waterholes in Tsavo East bringing many acres of good grazing into production during the dry seasons, while the target of fifty such places is no longer the remote pipedream it once was, but is now within the realms of possibility.

CHAPTER 19

Wiffle

ONE of the most endearing antelope, and also one of the most intelligent in the animal kingdom is the little dikdik. Weighing only eight pounds and standing barely one foot tall, they make up in grey matter what they lack in stature. This had been demonstrated to us first by Dicky during the short time he was with us, but was confirmed very forcibly by the three subsequent dikdik orphans we successfully reared and rehabilitated, the first of whom was Wiffle.

It is the habit of mother dikdik to conceal their young when very small, but Wiffle's mother had been rather unwise in the choice of a spot. She had selected a small clump of bush next to the local African school in Voi, so it was not long before some small boys discovered the baby and bore it triumphantly to their American teacher, who took it from them and passed it on to us.

She arrived in a cardboard box, a minute little 'Bambi' covered in short, greyish-brown grizzled fur with a longer, coarser, rufous-coloured crest on her forehead ending in an upright point of hair between the ears, which in our family soon came to be known as the 'frown', being a foolproof mood indicator. Normally this lay flat against her head, and was not very conspicuous, but it could be erected, each hair standing

228

straight on end to denote either alarm, irritation or discomfort. Wiffle's belly was pure white, her legs indescribably slender and delicate, her hoofs minute black, cloven points and her eyes large, liquid and expressive. But her most arresting and versatile feature was her nose, which was elongated and had fur growing right to the end of the nostrils. It could be moved from side to side, up or down, and like Dicky's constantly 'whiffled' this way and that, testing, examining, savouring and interpreting every faint scent wafted on the wind, opening up a kaleidoscope of hidden happenings of which we humans were quite unaware. It was the nose that prompted the name.

For several weeks Wiffle was prepared to spend most of the time lying peacefully in her box, and was brought out periodically for a feed and a bit of exercise. During her brief appearances, she very quickly learnt that her hoofs slid on the polished concrete floors of the house, and that it was wiser to stay on the carpet where possible. This did not deter her for long, though, and she soon mastered the floors, taking very short, careful steps. She learnt too that there was nothing to be gained by struggling when she lost her footing, and it was not long before by walking in this special way, she was able to venture off the carpet and explore the house.

The day came when she decided she had had enough of being shut in the box, and burst out of the top like a jack-in-the-box, landing on all four legs beside it with the 'frown' very much in evidence. We tried placing a heavy book on the lid, but the battering sounds that came from inside left us in no doubt that the infant stage as far as Wiffle was concerned had ended. Thereafter she had the run of the garden during the day and was brought into the house at night.

From a very early age, it was very evident that Wiffle was a 'one man' animal. I was the most important thing in her life, the only person permitted the liberty of picking her up and holding her in my arms without being flailed by needle-sharp hoofs; the only one from whom she would accept the bottle; the only voice worthy of response and the only being on whom she lavished all her attention, all her devotion and absolute loyalty. This was my reward for taking the place of her mother, and what more could any mother ask, even though it brought a host of problems. No other animal I have ever known offered such exclusive loyalty. She was almost like my shadow. I had only to look over my shoulder, no matter where I happened to be, and she was there, following along behind; and if she wasn't, I had but to call her name and she would appear as though by magic.

She was perfectly at home in both the garden and the house, and each day, full of the joy of living, she would race round and round, turning and jinking with lightning agility, darting under a shrub, leaping the terrace wall, dodging suddenly at right-angles to avoid collision with a chicken or the peacock, and inevitably ending up at my feet, panting with open mouth from the exertion, as though to say, 'How was that?', before beginning the game all over again. She particularly relished a bout of hide and seek with Angela, and would wait in ambush, sometimes in a kneeling position, to burst from cover at the appropriate moment and race off to another 'den', Angela trailing far behind in pursuit. At other times a game of 'dodge' was the order of the day, when Wiffle would stot a few paces on stiff legs, and when Angela dashed up, would merely bounce aside at an unexpected angle to avoid capture. Finally, when everyone had had their fill of games, she would tuck herself underneath a plant that offered a view of the door so that a close watch could be kept on my movements.

Wiffle had difficulty in climbing the slippery front steps, so every evening she had to be carried into the lounge where a selection of her favourite food plants were laid out on a newspaper alongside my chair, and carried later into the bedroom, providing sustenance for the night.

We had been in the habit of taking her out into the garden before retiring, to avoid an accident on the carpet, but this could involve a tedious wait of anything up to ten minutes, for she was extremely particular in the choice of a suitable spot. It had to be exactly where she had performed on a previous occasion, not necessarily always the same place though, for she had carefully established several little middens at various points in the garden, and the question of which one should be put to use each time seemed to require a lot of thought. The night excursions were not popular as far as I was concerned, for all sorts of strange sounds and smells were a constant distraction to Wiffle, and I felt the need to keep a very close watch on her lest prowlers like genet cats or jackals take a fancy to my pet, not to mention lions and leopards who might conceivably take a fancy to something a little larger! Finally I decided to make an attempt to house-train my charge.

This proved far easier than we had anticipated. A suitable tray was procured and filled with earth, and a fresh contribution taken from one of her middens and placed on top. After the evening feed, when it seemed likely from the size of Wiffle's tummy and her 'frown' that she must soon burst, she was plonked hopefully on the tray. Very carefully, she inspected every inch of it, paying particular attention to the contribution, and when she had satisfied herself that all was in order, she duly obliged, to a cry of triumph from all the spectators. The more 'seasoned' the tray became, the better she liked it, and so the tedious night vigils ended. Instead the tray was simply carried wherever it might be needed, and became an integral part of the household furniture.

Wiffle usually slept beside my bed, but one night she decided that she should be entitled to a place on top like the rest of us. She hurled herself vigorously against the mosquito net until we opened it up and allowed her in, but once inside she obviously felt cornered by the presence of the net, and so a compromise had to be made by tucking it in half way up the bed, so that she could sleep outside it on the bottom half of the bed, while I slept inside at the top. This arrangement worked very well for a time, and every night she would hop onto the end of my bed, cuddle up beside my legs, and settle down contentedly for the night.

After several weeks, however, she began casting covetous eyes on David's bed, which was much firmer than mine, having a bed board positioned beneath the mattress. It was almost a repetition of the Three Bears story that Angela loved: the floor was too hard, my bed too soft, but David's just right! After a thorough scrutiny and several trial runs, usually undertaken during the day when the bed was unoccupied, Wiffle made up her mind. David's bed it would be, and the struggle for possession that followed this decision caused a lot of amusement in the family. David, being a restless sleeper, was prone to lash out in his dreams, and heave and toss throughout the night until by dawn the bedclothes usually ended up round his neck screwed up like thin ropes. We were convinced that Wiffle would have some difficulty in remaining aboard, but she seemed grimly determined to stay put at all costs, and gallantly stuck it out, bouncing like a small ship on the ocean with every turn, and sometimes even being kicked clean off the end, only to reappear a moment later with frown erect signifying disapproval of this rough treatment. Despite these turbulent nights, for some extraordinary reason, she persisted in selecting David's bed each night, and would only resort to mine when David was in the throes of a particularly active dream!

To begin with, we thought that Wiffle's vocal capabilities were confined to the loud alarm whistle we had heard her wild counterparts make, and a softer version emitted as a warning when her nose shot downwards in a quick jerky movement. However, on a number of occasions, I had heard a soft, high twitter, often only barely audible, resembling the sound one normally associates with a bat, which is what, in fact, I thought it was. Angela, on the other hand, firmly declared that this noise was made by Wiffle, but at first nobody paid much attention, until one night we heard the sound repeated again and again, coming from under the bed. I switched on the torch, and sure enough, Angela was right. There was no doubt that Wiffle was the originator, and we have since learnt that this is a 'talking' noise which means in general terms either 'Hello' or 'Where are you?' or 'Watch me'. Later there was another surprise: we discovered that Wiffle could also scream. While quite prepared to amuse herself in the garden all morning, from teatime until bedtime she expected to be entertained, or

Feeding time at the orphans' nursery

Inseparable companions:
Stroppy, the rhino and Punda, the zebra

Lollipa, the buffalo and Stub, the rhino

Wiffle's third baby was a son

The changing face of Tsavo:
from dense bush country to open grassland

at least kept company. Almost on the stroke of 4 p.m. she appeared at the front steps (if I happened to be in the lounge or dining-room), or at the back steps (if I was in the kitchen, store or pantry), or outside the bedroom window if I was in there. She would open the conversation with the 'Where are you?' noise, and if I didn't hear, which was usually the case, she would repeat it a lot louder. If this still brought no response, she would give a loud scream, and there was no doubt about the interpretation of this sound. It meant, 'For the love of Pete, *come on!*' At first everyone rushed outside, fearing the worst, only to find instead an impatient little buck, eager for her afternoon walk.

These expeditions were the highlight of her day, for by now she was beginning to take an interest in activities beyond the confines of the garden. There were exciting smells to investigate and fascinating middens belonging to the wild dikdik. At first to be near one of these middens seemed to make her slightly uneasy, but as the days went by and no outraged midden owner sought revenge, she gradually became more confident and nonchalantly added her own contribution for good measure. She was always very careful to keep to cover where possible, and glanced skywards every now and then to make sure that there were no eagles in sight, for Wiffle, like most antelope, was endowed with an inborn knowledge of not only her natural enemies, but also her natural diet and behaviour pattern as well. We were certainly in no position to offer any instruction in such matters, but she knew them nevertheless. Her choice of food, for instance, in no way differed from that of wild dikdik, although there were one or two exotic variations, like biscuits, grain, an occasional lick of curry powder and the newspaper. One of her particular partialities were 'bobbles', the pealike buds of the wild hibiscus plants which grew in the bush, and the daily walk became known as 'bobbling'.

Wiffle always ate very rapidly for short periods at a time, gulping down copious quantities of a variety of plants without much regard to chewing, and then lying down peacefully to chew the cud at leisure. Obviously, in the wild state, it would be dangerous for an animal so small and vulnerable to have to be distracted for too long from the serious business of survival, and so the time needed to chew the food was left until later.

Cud chewing was a continual source of fascination to Angela, for the cud could be clearly seen rising up Wiffle's slender neck in a little round ball in response to a faint hiccup, was chewed dreamily for several seconds and then re-swallowed, its progress on the return journey again easily visible. Up and down the little ball would go, while Angela lay there entranced, watching the whole process in deep concentration.

As Wiffle's ability to cope with solids improved, so her daily milk requirements decreased, until by the time she was six months old, she was having only two feeds a day, morning and night, and from seven months, one at night was adequate, taken mostly from habit, I think. So much did she enjoy this bottle at night, that her back legs set up a kind of piston action: up, down; up, down; with ever increasing frequency as the contents dwindled, until by the time the very last drop had vanished, they seemed to have got completely out of control and even refused to stop, continuing to pump frantically for a full half minute or more! A very uncomfortable looking Wiffle, with barrel-like sides, would then head determinedly for the earth tray, following which she looked, and no doubt felt, a lot better.

One extremely important prelude to the daily bottle, which dated back to the very beginning, was the metal watch strap on my wrist, which had to be nuzzled first. If this formality was overlooked, then the bottle was refused and although the significance of this habit defeated us, it was nevertheless of the utmost importance at mealtimes.

Wiffle, in the light of the subsequent dikdik we have kept, was a particularly timid little animal and, understandably, the larger orphans were a source of constant upset, for she saw them only infrequently when they happened to trespass into the garden, and snatch a few delicacies. The column would be headed by Eleanor, with the other elephant following close

behind, then Rufus, the buffalo and 'tail-end Charlie', good
old Stub, plodding drearily along in the far distance completely
switched off! Wiffle's first glimpse of this cavalcade must have
been a most unpleasant experience, for she was caught quite
unawares engrossed in a game with the peacock. Suddenly
Eleanor appeared round the hedge followed by her satellites.
Wiffle stopped dead, her eyes almost starting out of her head
while her fringe sprung up and her nose shot down emitting a
terrified nose whistle. The next moment she had vanished in a
series of enormous bounds, and when we finally unearthed her,
crouching beneath a thorn bush at the back of the house, she
was still trembling violently. Thereafter, of course, she had
plenty of opportunities to view the other orphans at longer
range, but even so she never managed to get used to them.
Lollipa the buffalo, particularly terrorized Wiffle, for she had a
habit of playing truant and leading the herd boy a dance by
thundering round and round the garden when he was trying
to round her up again. On such occasions, we could be sure
of a 'Wiffle hunt' later on.

Nocturnal visitors around the house were another source of
bother, for while we humans slept peacefully, oblivious to any
happenings outside, Wiffle was constantly alert, conscious of
every movement and sound. Her eyes never closed completely,
even when she slept, and her ears turned constantly with every
waking moment, while her nose whiffled all the time. Every
now and then she would stand up on the bed to stare out
of the window into the night. Her muscles would tense at
any sudden sound, and when a huge bull giraffe decided to
prune the melia tree on the lawn one night, standing just outside our bedroom window making vulgar crunching and gulping noises, Wiffle's nerve broke, and she plunged under the
bed.

Although we didn't really begrudge the giraffe some foliage
from our tree, we felt obliged to intervene for the sake of
Redhead, a wild redheaded weaver bird, who had nested in
that particular spot for many years, and who, we felt, had more
right to it than the giraffe. Redhead's current nest hung pre-

cariously on the end of a thin twig easily within giraffe reach, so I nudged David, who leapt out of bed and stared wildly around, obviously still absorbed in a James Bond-like dream. I urged him to take action, and when he had managed to sort out fact from fantasy, he seized the torch, muttering with displeasure, and burst out of the front door, while I was an interested spectator from the bedroom window. The giraffe seemed to pause a moment, gather all four enormous legs beneath him, lower his neck, and launch himself with a thundering of hoofs at the pergola dividing the carpark and garden. It was very obvious to us in those few brief seconds, even by moonlight, that the opening in the pergola was certainly not sufficiently large to accommodate him. His outstretched neck passed through with ease, then his body obliterated the skyline; there was a scuffling, a creaking and groaning of timbers, and to our amazement the entire pergola rose clean out of the ground to be dumped down again, intact, further on, while the giraffe galloped off down the hill!

Back in the bedroom, no amount of reassurance on our part could persuade Wiffle to return to the comfort of the bed, so she spent the remainder of the night underneath it.

For a long time we had been grappling with the problem of what to do with Wiffle should we have to go away at any time. We now had to face this issue, for Angela was due to begin boarding school, and we had to take her up to Nairobi, involving an absence of about four days. Only I could pick Wiffle up, and no one else had ever succeeded in feeding her either, although, fortunately, the bottle was no longer of such importance for she was practically fully grown. The greatest hurdle would be getting her into the house for the night, for she could not manage the front steps unaided. Finally we decided that Jill would have to remain at home to cope as best she could, so she was briefed in great detail and left with the all important watch strap.

When we returned, having deposited Angela at school, we found Jill, amidst clouds of smoke, burning herself an egg for supper, while Wiffle was standing gloomily in a corner with the frown in evidence. As I entered the door she looked up eagerly, and the expression of joy on that little face had to be seen to be believed. I picked her up and carried her back into the lounge, conscious of the excited beat of her heart, and when I put her down, I noticed that she had shed copious quantities of hair. We concluded that shock must have caused this sudden moult, and David recalled having noticed the same thing in animals that had been shot. While Jill recounted all that had taken place in our absence, Wiffle proceeded to behave in a most extraordinary way, so overjoyed was she to see me again. She crouched at my feet, then jumped up and tried to walk on her hind legs, leapt onto my lap and then hopped off again, hurried round the coffee table and repeated all these antics over again, until we began to wonder if the reunion had deranged her temporarily.

Apparently, as we had feared, Jill had experienced difficulty in getting Wiffle into the house at night, and all attempts to catch her had ended in failure, for although she would allow Jill to approach within scratching distance, any attempt to seize her was artfully sidestepped by a quick bound sideways. By dusk Jill was becoming desperate, and had enlisted the help of the neighbouring scientists. A lot of time was wasted coaxing and pleading, all in vain, and just as everyone was at the point of despair, and beginning to accept the fact that Wiffle would have to brave the night outside, she launched herself at the front steps and scrambled up, much to everybody's great relief.

Having heard all the news I decided to take a bath, and Wiffle, bent on keeping tabs on me at all costs, came along as well. Relaxing in the bath, I momentarily disappeared from Wiffle's view, and the next moment she leapt clean over the side of the bath and landed on top of me, causing us both to almost jump out of our skins. The chaos that followed can well be imagined, with Wiffle thrashing around wildly, and me struggling to get a hold of her before she succeeded in drowning herself. It was several seconds before I managed to lift her out; a very bedraggled, shaken, and subdued little animal, who shivered and sulked for hours afterwards. Thereafter the bathroom was shunned, and she contented herself with an occasional peep round the door whenever I happened to be inside.

One day the conversation piece was the desirability of procuring a suitable spouse for Wiffle, and I had taken the opportunity of a stay in Nairobi to visit the Nairobi Park Animal Orphanage with a view to looking over any likely suitors. However, to my disappointment, there was only one adult male there at the time, and he was already attached, so the prospects were not hopeful. I decided that Wiffle must be rather lonely, and wondered if perhaps a young goat might prove a good companion until such time as we could acquire a mate for her, but this suggestion was firmly vetoed by David, who declared that under no circumstances whatsoever would he agree to having a goat on his bed as well as a dikdik! Instead we discussed the possibility of introducing Wiffle to the car, so that we could take her down to the airfield, where a very cheeky little male dikdik lived, who had, on several occasions, even dared to challenge the car in defence of his territory, but our attempts to precipitate an introduction always ended in failure, for the wild dikdik never put in an appearance on such occasions. Little did I realize at the time that I might have spared myself all this anxiety for, when the time was ripe, Wiffle proved quite capable of selecting her own spouse without any help from me. But we were not to know this at that time.

We had, on one occasion, experienced what could be described only as an intruder, for he couldn't really qualify to the title of burglar having taken only a tin opener and a tablecloth, and helped himself to a beer and some fruit salad and raw sausages out of the fridge. Perhaps 'uninvited guest' would be more appropriate, and through all this, we had slept peacefully and never heard a sound. We never did have the pleasure of seeing this gentleman, either, for although his tracks were followed the next day, they became obliterated by those of a herd of buffalo, so we took comfort in the hope that perhaps he had actually encountered the buffalo on his travels and that this might act as a good deterrent to a return visit.

It was therefore only reasonable that this episode should

spring very much to mind, when we were suddenly aroused one night by the sound of splintering glass from the direction of the lounge, and the simultaneous reaction of Wiffle, who again plunged underneath the bed.

Scrambling out of bed, David crept stealthily towards the lounge, muttering something about catching the blighter red-handed, while I groped for the torch, and followed with a thudding heart. I arrived just in time to hear David fling open the front door, at the same time leaping over a pile of broken glass lying on the floor, and erupting onto the front verandah all set to grapple with the intruder. There followed instead a startled exclamation and a series of loud, unexpected snorts, as he collided with a rhinoceros! Rufus had somehow got out of his stable, and had come to 'knock us up', thrusting his horn through the french door leading into the lounge. He was obviously very pleased with himself at having extracted such a prompt response to his summons, and was plainly all set for some fun. Hurriedly, David pushed past him to entice him out rather than in, and having succeeded in doing this, Rufus could contain his exuberance no longer. Tossing his head furiously, snorting and kicking up his hind legs, he proceeded to career around the garden, returning at frequent intervals to make short rushes at David, who was forced to take evasive action just like a bullfighter, holding his *kikoi* (the popular night attire for men in Kenya, consisting of a gaily coloured loin cloth wrapped round the waist) as a decoy. Gradually, the 'game' was steered towards the stable, but it was a good half hour or more before we regained control of the situation and managed to inveigle Rufus back into his stable, and by the time we finally returned to the bedroom we both felt completely exhausted.

Poor little Wiffle, in extreme agitation, was still indulging in frantic nose jerks and whistles, and no doubt had been under the impression that 'her mother' had been involved in a fight to the death outside, if the noises were anything to go by! From under the bed the odds must have sounded heavily in favour of the enemy. As I tried to comfort her I was very conscious of the terror she must have been subjected to, and the relief that was now so obviously flooding over her as she licked my face.

When Wiffle was eight months old she was, to all intents and purposes, mature, and I noticed that the afternoon 'bobbling' sessions took on even greater significance. Now the two small glands situated below her eyes, which are found in both the male and female of the species, were obviously active, and exuded a tarlike substance. She would go out of her way to rub them on the ends of any small twigs protruding from a shrub of suitable size, or on the tips of coarse grasses, leaving each time a small deposit of the tarlike substance. Many of the twigs and grasses selected for this purpose were already tipped with small balls of this glandular secretion, some the size of a pea, which had been deposited there by the wild dikdik. Such 'signposts' were eagerly scrutinized and carefully smelt before Wiffle added her own contribution to the little ball. I also observed that when the glands were moist a certain small species of fly seemed to be attracted by the secretion, and several of these insects persisted in clustering around each gland causing Wiffle a good deal of irritation.

I had been under the impression that Wiffle was far too attached to me to ever feel the desire to seek the company of her wild brethren, but in this I was wrong, and Nature's dictates proved the more compelling. She took to disappearing for the odd hour or two each day, but was always at the front steps for her daily walk. And then, one day she wasn't. Frantically, I searched the garden and its surroundings but there was no response to my calls. When dusk faded into the night I was extremely anxious, imagining all the most gruesome possibilities that could account for her absence, convinced that she must have come to grief. Throughout the night I was up and down like a yo-yo to see if she had returned. To my immense relief, however, she was at the front steps to greet me first thing in the morning, but I could detect a certain restlessness in her, and she kept gazing beyond the garden into the bush below the offices. That afternoon, she again vanished, so I decided to take a stroll to try to solve the mystery of where she went to on such occasions, for although I was fairly certain that the attraction must be a boyfriend, I was curious to know the location of his territory so that I would know where to look for Wiffle in future.

I came across many wild dikdik during the course of my wanderings, all of which bounded away into cover, until finally I happened upon a little group of three, who were watching my progress with interest. Two darted off as I approached, and one raced joyfully up to me to greet me with habitual affection.

We returned home together, Wiffle and I, and I felt happy to think that this little animal I had raised, who had lived such a secluded and isolated existence in an unnatural environment, had nevertheless succeeded in establishing contact with her own kind. At the same time, I couldn't help feeling sorry for Wiffle, torn between two loyalties, and I hoped for her sake that she had made the right choice.

For some weeks she continued to keep the rendezvous below the office, and could usually be found there each afternoon, until one day she and all the others vanished and were nowhere to be seen. Finally, several days later, news came that she was in Phil Glover's garden, and as his house was identical to ours, I assumed that she must have become confused, and turned up in the wrong garden accidentally. I therefore set out immediately in the car to retrieve her, but no sooner had we arrived back home, than she began looking intently in the direction from whence we had come, and several hours later had again disappeared from my garden, and reappeared in Phil's. Once again, I retrieved her and brought her back, but realized that no accident was responsible for her choice, for I intercepted her soon afterwards heading back yet again. As I, too, happened to be going that way, I stopped the car and gave her a lift, depositing her with apologies to Phil and his wife, at their house.

Barbara Glover was a very keen gardener, and did not exactly relish Wiffle's presence, for she found her plants being pruned rather ruthlessly, and although I provided biscuits and grain to try and take the edge off my former pet's appetite, the ends of all the most treasured plants continued to be neatly nipped off.

It was apparent that Wiffle had fallen for a male dikdik who already had a wife, which was a bit disconcerting in view of the fact that dikdik are reputed to mate for life. Nevertheless, Wiffle presenting the eternal triangle was constantly seen in the company of this male and his other wife; and what is more, provided ample evidence of the male's infidelity, for she grew more and more portly as the months progressed. Every time I went over to see her, and picked her up for a quick hug, I could detect a distinct difference in her girth. There was no doubt about it. She was pregnant.

Unfortunately the change was not confined only to her figure, for Barbara Glover complained of being attacked and 'bitten' whenever she ventured out of her own house, which certainly added insult to injury! I was incredulous of this allegation for I had never heard of anyone being 'bitten' by a dikdik, and couldn't begin to visualize how Wiffle could possibly achieve this anyway. In support of her claim, Barbara pointed to her ankles, and I had to concede that these were covered in thin scratches. Phil also testified that these wounds were inflicted by my dikdik, so I had to accept that it was so, although I was baffled by this turn of events, for, quite apart from anything else, Wiffle had always been rather nervous of strangers and of a timid temperament. Finally, Barbara Glover had to resort to donning a stout pair of gumboots whenever she ventured into her garden, and I had to agree that she had a genuine grievance, for wearing gumboots in the heat of Voi was certainly no joke, but there wasn't much I could do about it.

Eventually conclusive proof was provided when I actually witnessed this extraordinary behaviour for myself. Wiffle would dart into the attack, quick as a flash, her head down and frown erect, and with a lightning upward flick of her head, scratch Barbara's leg with her lower incisors, leaving a thin, but nevertheless, quite deep, scratch.

I suppose that this aggressive behaviour was connected with a desire to defend either her boyfriend, or her territory, or both, but it seemed to be aimed only at Barbara to begin with. Possibly, in Barbara, she recognized a contestant for the garden!

The gestation period for dikdiks is six months, and towards the end of this time Wiffle literally waddled along, the discomfort caused by the advanced state of her pregnancy evident in her lethargic movements. Her little udder was hard and distended, and we knew that the baby would arrive at any moment. Sure enough, a few days later, a message came from the Glovers – Wiffle had regained her sylph-like appearance, and was once again playful and active. The baby must have been born, but although we all searched the garden and the bush round the back of the house, Wiffle strolling nonchalantly behind, we were not going to be allowed the privilege of being shown it, and failed to discover where she had hidden it. Not until six weeks later did she bring it out for all to see, and after that it accompanied her most of the time and became remarkably tame. It turned out to be a female, and while it was not possible to actually handle it, one could approach to within a few paces before it slowly walked away. In doing so, however, one stood to incur the displeasure of Wiffle, who was quick to resort to biting in defence of her baby, and even I was not immune on such occasions.

Some months after the arrival of her baby, Wiffle stepped into the role of principal wife to the male, whose other wife mysteriously vanished, probably having fallen prey to some cat or eagle. She also acted as guardian to the first wife's orphaned youngster, which was roughly the same age as her own, and all three of them could be regularly seen wandering around Barbara's garden most afternoons, while father watched anxiously from a safer distance below the terrace.

When the youngsters were about six months old, they became more independent and frequently disappeared from the garden, probably busy searching for their own mates and new territories, and by this time, Wiffle was again taking on a matronly look. Exactly three days short of seven months after the arrival of her firstborn, (206 days, to be precise), her second daughter was born, and this time, we were fortunate in being able to have a peep at the new baby before she removed it from near the garage, where it had been born, and concealed it properly in the bush behind the house. Again, some six weeks passed before it began to accompany her most of the time.

During all this time I had been busy rearing another orphaned female dikdik, known as Winkle, whose mother had been run over on the main road. I had also acquired a baby impala, so Winkle was not turned into the household pet that Wiffle had been, for she had to keep the little impala company at all times, otherwise it became frantic and rushed round the garden bleating pitifully.

Winkle demonstrated very forcibly, yet again, the individuality of animals of the same species. Where Wiffle had been timid and retiring, Winkle was bold and inquisitive; where Wiffle had been totally dependent on me, Winkle was independent and adventurous. Her supreme confidence was fortunate in that it had a calming influence on Bunty, the impala, who, although gentle and affectionate, lacked the independence and assurance of dikdiks. In fact, she could best be described as plain 'jumpy', and was prone to do the most stupid things, taking fright at her own shadow. Winkle, on the other hand, was all for investigating anything unusual, and even the sudden arrival of Eleanor and her followers did not daunt her. She simply pretended she had failed to notice them, edging closer all the while to satisfy her boundless curiosity. Only the ominous shape of a Martial or Bateleur eagle overhead sent her darting for cover.

Both Winkle and Bunty loved their afternoon walk, and would follow wherever I chose to take them. What Bunty lacked in intelligence, she amply made up in beauty and sheer grace of movement. With no apparent effort, she would leap fluidly over the bird bath or a bush, or stot on stiff legs with a rocking motion, seesawing from her front legs to her hind legs as though propelled by a tightly coiled spring that sent her bouncing high into the air at each stride.

Impala must surely be one of the most beautiful of antelope; the dull red colour of the Tsavo earth, with black socks, black eartips and a pure white rump, two black stripes on either side of the tail, and another running down the centre of the back and ending at the tip of the tail itself. The males have sweeping horns that spread out and converge again slightly at the ends. Gregarious by nature, having a social structure in which a dominant male lords it over a harem of ewes, while the other males keep their distance in bachelor groups until such time as a contestant for the harem plucks up sufficient courage to challenge the old man, we realized that Bunty's dependence on Winkle was only natural, but we anticipated difficulties because of it when the time came for Winkle to go courting.

This day came when they were both six months old, and so distraught was Bunty during Winkle's absences that the children and I had to take it in turns to sit with her in the garden. However, happily, for some weeks, Winkle returned each morning and confined her outings to the afternoons and nights, so Bunty finally realized that these separations were not permanent.

I had a sneaking suspicion that Winkle was following in Wiffle's footsteps by casting eyes on Barbara Glover's garden; a development I knew would not be popular. These fears were confirmed during one of Winkle's more protracted sprees when she had spent several nights away from home, and suddenly turned up in the house next door to the Glovers', fraternizing with some rather puzzled lesser kudu – obviously the nearest thing to an impala in the vicinity. I set out at once with Bunty to entice her home, and our route took us through Barbara's garden, where we encountered the Wiffle family. I was very interested to notice that Bunty paid not the slightest attention to any of them, and could quite obviously differentiate between individuals, for when we found Winkle next door, the reunion was touchingly joyful. Winkle bounded up to Bunty at the same time as Bunty rushed up to Winkle. Standing on hind-legs, with forelegs resting against Bunty's shoulder, Winkle rubbed noses with Bunty for several seconds, and then proceeded to dance round playfully, until she suddenly seemed to realize that I had been neglected, and switched her attention to me. We all then headed homewards, but when we encroached on Wiffle's precincts, I had to carry Winkle, for Wiffle advanced with frown erect and in a belligerent manner.

Walks of this nature to locate Winkle became almost a routine, and were very popular with Bunty. We never failed to find Winkle, and more often than not I would happen to glance around to find Winkle and Bunty greeting one another, Winkle having appeared silently from the bush behind. We knew exactly where to find her too – just short of Barbara Glover's garden.

One day Barbara announced casually that Wiffle had disappeared and had not been seen for some time. I was a little anxious at this news, but happily my fears were unfounded, for Wiffle turned up in the next door garden, belonging to Tim and Mel Corfield, where she was accorded a warm welcome and given biscuits and grain. She even took possession of their house, walking on the polished floors in the same way as when she was very small, but, again, unfortunately, she felt the need to demonstrate her right to be there by periodically 'biting' both Tim and Mel, usually when they returned after being absent for a few days. When she was particularly belligerent, they had to shut her in the house while they were in the garden, or shut her out of the house when they were inside. Father Wiffle was also in evidence, although he kept his distance when Tim and Mel were at home.

When Wiffle moved territories she was again pregnant, and this time she gave birth to a son only 175 days after the birth of her second baby, which was short of the normal gestation period of 6 months. This third baby was accorded rather different treatment from the other two, and appeared to be allowed to accompany his mother frequently throughout the day even when only a few days old. It was also a particularly playful baby, and kept Tim and Mel amused with its antics for hours on end.

Meanwhile, I too, had acquired a little male dikdik which we called Dudu, who resembled Wiffle in temperament. His arrival cut short our visits to Winkle, for Bunty was delighted with the newcomer, and preferred to stay at home with him. They even slept together at night, until one morning, when Dudu was three months old, I opened the door of the stable to find, to my horror, that he had a broken foreleg; fractured in two places below the knee with a piece of bone detached in between. How this happened remains a mystery, but we concluded that Bunty must have inadvertently stood on him during the night.

David looked at the leg and decided that he would not attempt to repair it himself, but would fly Dudu up to Nairobi where the leg could be set by a vet, for to a dikdik, sound legs are all important to survival. Matters were complicated further by the fact that it was a Sunday, but we succeeded in making an appointment by phone, and set off as soon as possible.

Dudu behaved very well during the flight, displaying the habitual intelligence of his kind. He lay quietly in his box and did not seem particularly concerned by the strange surroundings in which he found himself, or the noise of the plane. The wait in the surgery was rather an ordeal though, for there was an assortment of other patients present, mostly dogs, ranging from very large to very small, and the barking, growling and howling that went on must have been terrifying for poor little Dudu.

But, when his turn came, he accepted his lot philosophically, although I held him while he was given the tranquillizer, but he behaved extremely well throughout, and after it was all over, began to nibble unconcernedly at the lucerne in his box.

Dudu learnt to manage admirably with his stiff leg, and soon mastered the art of manipulating it. Unfortunately, however, it became apparent that the plaster had been put on too tight, for the little hoof protruding from it began to swell and sweat until it was four times its normal size. We tried cutting the plaster and opening it up slightly, but even this did not help, for, unknown to us, a small piece underneath was still restricting the circulation. Finally, a visiting vet passed by, and after taking one look at the foot, announced to our dismay that matters were extremely serious and that it might even be already too late to save the leg. In any case, the plaster would have to be removed immediately.

We had no anaesthetic, and so we had to perform the operation without. With the help of several scientist volunteers, we forcibly held Dudu down on a table while the plaster was being cut away to reveal the fracture, and we were horrified to find that there was no sign of the bone having knit at all, although the leg had been in plaster now for three weeks. The vet was gloomy about the prospects of its mending, and hazarded only a one per cent chance of success. He thought it more likely that the leg would turn gangrenous and fall off, which was an appalling thought. Nevertheless, he reset the bone and replastered the entire leg, this time more loosely with a lot more cotton wool, and Dudu was at last released. He bounded off with his frown on end, and began to feed, none the worse for the ordeal.

Eight weeks later we removed the plaster, not knowing quite what we would find underneath. Imagine our joy when we discovered that the leg had healed completely and was strong and straight. Today, it is as good as new, and presents no handicap whatsoever. Dudu runs, plays, jumps and fights just like any other male dikdik, and the only means of telling which leg was broken is by the presence of a faint thickness where the calluses have formed.

Dudu is now seven months old and beginning to think about a wife. As it is the males that seem to be territorial in dikdik society, we are optimistic that he will not feel the inclination to end up in Barbara Glover's garden like the other two, but will stay instead in ours, and may perhaps even entice a mate into it. A lot of time is devoted to rubbing his horns up and down upright stems and sticks, sharpening them against the day when they must be put to use, and a lot of time is spent touring the limits of his territory, examining any evidence of callers, and carefully affixing little globules of glandular secretion at various points. He enjoys taking a look at other territories too, and occasionally disappears for several hours at a time, or takes advantage of a walk with Bunty and me to explore further afield. However, this practice once led to reprisals by one wild male, who chased poor Dudu right up to the front steps, and was so intent on revenge that he persisted even when Dudu sheltered behind me, until he suddenly came to and realized where he was going!

CHAPTER 20

The Tragedy of Rufus

WHEN one raises an animal, particularly any of the big five that are normally labelled 'dangerous', one is left with a responsibility for it that has a habit of lingering long after the animal itself. Even in the case of Houdini, the rhino who ended up in the Teita Reserve, it was whispered that this animal had been tamed by the National Park in order to drive people away from the areas near the boundary. And a Union official listed amongst his grievances for good measure, the fact that *Saa Nani* (the name given to David by the Africans, meaning, in Swahili, two o'clock, because it was usually not until two o'clock that they were allowed off for lunch!) thought more of the animals than of people, which, in some respects, was probably correct, for wild animals were David's life, and his first thought in most situations was what was in the Park's best interests, and not what happened to be expedient at the time.

Our first elephant, Samson, who had been wild for at least five years, was chosen as a likely candidate to shoulder the blame for raiding the prison premises near the Manyani entrance to the Park, although, in actual fact, he was quite innocent and was known to be forty miles away at the time. The true culprits were several old bulls who had grown bold and blasé about both traffic and people, and who frequented the area around the camp perimeter, and soon learnt that forbidden fruits were sweeter, marching right into the prison compound in search of water and the vegetables that were sometimes grown there. They had even been known to eat baskets of fresh vegetables lying at the railway station awaiting collection. Nevertheless, it was of Samson that the official thought, when he sat down to lodge a complaint, and we received a letter saying, 'We have heard rumours that this particular elephant known as Samson was formerly tamed by National Parks, and when he proved himself to be nuisance, was then chased away and thus the reason why he is too familiar with human habitation.' The writer then went on to say that while he was aware of the benefit to the country of wild animals, he was afraid that 'for Samson special segregation should be done', for he was 'too clever to be of wild behaviour.' The letter enlarged on this statement by saying, 'when he wants to cross the trench and he sees that it is too deep, he fills the place he intends to pass through with loose soil, and then passes without any difficulty. Also he can crawl and even jump' (which were certainly not amongst Samson's former accomplishments!). 'When he breaks water pipe and he is disturbed, he becomes irritated and takes away taps with him.' And to summarize the unusual cunningness attributed to this animal the letter ends up by declaring that 'Samson is wonderfully serious creature and he is more than what I can describe'. The writer asks that 'Mr Sheldrick, who is directly concerned with the taming of wild animals', kindly do something about it!

The first thing we had to do was to convince the prisons that the elephant in question was not Samson, and some Rangers armed with thunder flashes and Very pistols were despatched to try and discourage the true offenders from visiting the prison compound.

There is always an element of risk attached to the keeping of large animals, particularly rhino and buffalo, for even the most docile has its moments of irritation when provoked, and once the balance of power tips in the animal's favour, it automatically becomes a potential danger. It was for this very reason that as soon as Samson, or any one of our other elephant, showed signs of becoming difficult to manage, we urged them to sever their human connections completely. We had tried to do the same for Ruedi, but until now Rufus, always gentle, affectionate and extremely docile, more a human than a rhino, had never given us any cause to consider taking this step.

And then, one day, an incident occurred that ended in a double tragedy – the death of the old Turkana tribesman who herded the elephants, and the ultimate death of Rufus himself, after he had been with us for ten years.

The tragedy was an accident really, for Rufus had not intended any harm to the old Turkana, of whom he was, in fact, very fond. Instead, he had become incensed at another man, who had been detailed to accompany the orphans on this particular day in the absence of the normal second attendant. We suspected that Rufus had good reason to lose his temper, for we heard later that this particular man had been seen beating him that morning, and very likely had been doing the same that afternoon, although he feigned innocence. At about 5 p.m. Rufus turned on him. His cries for help as he dodged brought the old Turkana at the run, and Rufus, who was by now very excited, swung round at his approach and in so doing happened to drive his horn deep into the old man's thigh.

The first thing we knew about it was when the man who had been the target appeared at the house and blurted out the news that the Turkana had been killed by the rhino. David immediately rushed to the place where he found the Turkana squatting on the ground holding his leg in an attempt to staunch the flow of blood. Although the wound was deep, no bones had been broken, so the wound was bound up and the old man taken straight to the local district hospital in Voi. Most unfortunately, the doctor was away, but David saw the dresser in charge clean and stitch the wound and make the old man as comfortable as possible, before leaving him to go and look for Rufus and the other orphans, who had been left to their own devices.

I, meanwhile, was slightly alarmed to see Rufus heading for the house on his own. Knowing no more than what I had heard, that he had just killed the Turkana, I wondered how he would behave towards me, for I knew that I would have to intercept him in the garden in order to prevent him from coming right into the house. I was relieved, therefore, to be greeted by the same calm, sloppy old Rufus, who nuzzled me gently to see what I had to give him, before walking to the garden tap, turning it with his horn, and helping himself to a drink. I then took him up to his stable and locked him inside.

Long afterwards David returned, looking agitated.

'We can't find Rufus anywhere,' he said. 'They say that he has gone berserk and charges everyone on sight, but he seems to have vanished into thin air,' he added.

'But, he's in his stable,' I replied, astonished. Now it was David's turn to look surprised, for he and the Rangers had been scouring the bush for the last two hours!

'How the devil did he get in there?' he asked, and I then recounted what had happened.

The reports that Rufus had gone berserk, were not entirely without foundation, however, for while we could handle him with complete safety, he behaved very differently whenever an African approached. We thought that this was only a temporary development in view of what had taken place, and that he would probably calm down later.

That night, we burnt the midnight oil discussing Rufus' future, for it was quite obvious that he could no longer be kept at the headquarters. We discussed zoos and the orphanage in Nairobi, and finally concluded that having had freedom for so long, he would not be happy in either. We therefore resolved to attempt to rehabilitate him, and decided to move him to Aruba, where we could keep an eye on him, and supplement his diet until such time as the rains broke. The Lodge itself was surrounded by an elephant-proof ditch, which would ensure that he did not have access to the compound, but at the same time would be able to feel that he was near humans.

The next morning came the news that the old Turkana had died. This came as a complete bombshell, for the wound had not appeared to be of a serious nature, and certainly did not look as though it could possibly prove lethal. A postmortem examination, carried out by the doctor on his return at our request, revealed that one of the arteries had been nicked, and the fact that the old man had been slowly bleeding to death had not been noticed during the night.

So, Rufus had killed a man! It seemed unbelievable, for out of all our orphans, it was he that allowed little Angela to clamber all over him and ride on his back; to crawl underneath him; to swing on his horn. Even so, it was a very different Rufus that came out of his stable that morning. A rhino bent on revenge, who charged any African that came within range in a most determined and aggressive way. Nor could we persuade him to go back inside his stable either. He had learnt that he could get the better of frail humans, and overnight had become unmanageable.

In the end, one of the Rangers volunteered to act as live 'bait', and ran inside the stable. Rufus thought the moment of retribution had arrived, and as he followed in hot pursuit, the Ranger nimbly vaulted over the partition dividing the two stables in the nick of time, while David slammed the door.

We had problems also in trying to entice him into the travelling crate. No amount of coaxing helped, and in the end we had to resort to feeding him a doped mango. This had the desired effect, and within ten minutes his head drooped and his eyes closed, and by pushing him gently forward, he sleep-walked straight into the box.

By the time he had arrived at Aruba, the effects of the drug were beginning to wear off, and he woke up to step cautiously out of the box and begin to explore his strange surroundings. Two Rangers kept watch over him that night, from a safe distance, in case he was still not yet fully alert, and the next day, having satisfied themselves that he was behaving quite normally they left him to fend for himself. For three days Rufus was seen near the Lodge and appeared to be quite happy, but then he disappeared, and although we hunted for him, we failed to locate him. Ten days later, while driving back from the Ndara bore-hole, the Assistant Warden spotted vultures. He went over to investigate, and lying there, his throat and body lacerated by deep gashes, lay poor old Rufus, victim of a vicious attack by lions, who had failed to kill him outright and had left him so severely injured that he had succumbed.

The death of Rufus was felt very deeply by everyone in Tsavo East, and also by hundreds of people all over the world who had made his acquaintance and had been charmed by his gentleness. One old lady, resident in Spain, made a point of visiting the Park every year for the express purpose of seeing Rufus. We had all loved him and the suffering we knew he must have gone through made matters very much worse. Pampered and petted all his life, he must have been very puzzled at being suddenly ostracized, and because he was quite accustomed to being expected to consort with a variety of animals, he had in all likelihood wandered straight up to the lions, probably glad of some company, and had paid dearly for this folly.

Another problem was beginning to loom up, posed by Lollipa the buffalo, inseparable companion to Stub, who was now rising five years old. Belonging to a species that matures a lot quicker than rhino, Lollipa had overtaken Stub and was virtually an adult when Stub was still a calf, although they were both the same age.

Like all the buffalo we had ever kept, she enjoyed chasing children, and anyone else who betrayed obvious fear. But, she was also the most outstanding of our long list of buffalo orphans, and she never forgot the part played by Mavis Hucks and myself in her early life, singling us out of even a crowd of many people for preferential treatment. She never missed an opportunity of going out of her way to greet us, and would often dodge the orphan attendant to do so. Angela, who was one of the principal chasing targets, grew very adept at shinning up the nearest tree in the garden whenever Lollipa arrived unexpectedly. The labour, and in particular their wives and children, were also very much on their guard when she was in the vicinity, being another of her quarries. It was becoming very plain that we would have to start making plans for her future.

Our eastern neighbours, the Galana Ranch, who in addition to ranching cattle, carried out experiments aimed at assessing comparisons in growth between game animals and domestic stock, had already gathered together the nucleus of a small buffalo herd, which was to be used for this purpose, and had expressed a willingness to give Lollipa, and any other buffalo we wished to dispose of, a home. So, having extracted a solemn undertaking that none of our orphans would end up in a slaughter house, or be subjected to experiments in the name of science which might cause them unnecessary discomfort, preparations were put in hand to move all our buffalo, including Lollipa, as soon as possible.

For sentimental reasons mainly, and because of Stub's attachment to her, Lollipa was left until the last, but her departure was delayed even further by a serious misfortune that overcame her down by the Voi River where the orphans went to feed. They had stumbled on a wild rhino who had been taking a nap in the grass, and Lollipa's liking for rhino on this occasion proved her undoing. The rhino promptly charged her and tossed her twice, goring her in the foreleg and in the stomach, before going on its way.

Being the only person who could handle Lollipa, I was hurriedly summoned by the Ranger on duty and arrived on the scene to find a very dejected buffalo standing disconsolately with her head hanging down, and fluid from the abdominal cavity dripping from the deep wound in her navel. Standing close beside her in dumb but touching sympathy was Stub, quite obviously aware of Lollipa's plight. As I called Lollipa, a flicker of recognition came into her eyes, and she permitted me to clamber underneath her stomach and assess the damage, flinching with pain as I prodded her stomach in an attempt to determine the extent of the injury. I decided finally that neither wound was very serious, and could do no more than bathe them both with healing oil, and walk Lollipa slowly back to the headquarters, pausing to rest at frequent intervals. All the way, Stub walked close beside her instead of lagging behind as was usual, and one could sense the bond that existed between these two animals and the sympathy that one had for the other. I realized then with great sadness just how much they would miss each other when the day of parting came.

Wild animals are endowed by Nature with phenomenal powers of healing and recovery and Lollipa proved no exception. Only a slight lethargy and loss of appetite for a period of a week betrayed the fact that all was not quite as it should be, and within two weeks she was back on form, sending the labour scuttling for the nearest tree on her way past the lines. And then, it was her turn to join the Galana buffalo herd, and as the lorry drove off, with Lollipa aboard, we were sorry to see her go. Sorrier still was poor Stub, who wandered aimlessly around the stockades calling pathetically, refusing to accompany Eleanor and the elephants down to the river as before. This misery gave way to aggression as Stub resisted any form of control, bent on going her own way in the solitary rather lonely fashion of her kind, until one day she discovered Barbara Glover's garden, and forced us to have to start thinking again.

Finally, following a suggestion by the Director, she was moved to the Nairobi Park Animal Orphanage to make the acquaintance of their male rhino, Stompy, with a view to eventual rehabilitation as a pair in Mount Elgon National Park, where rhino used to occur in the old days. We thought this as good a suggestion as any, and felt that Stub would stand a better chance in Mount Elgon rather than risk the fate of Rufus.

Some weeks later, I had an opportunity to visit her and see for myself how the friendship was progressing. To my surprise Stub was planted fair and square in the middle of the enclosure, while Stompy, the rightful owner who was very much larger, cowered in the perimeter ditch, every now and then being warned against taking liberties by a series of aggressive snorts directed in his direction. I couldn't help feeling rather sorry for him, but I might have realized that this situation would not be borne indefinitely, and that Stompy would attempt a come-back. This he did, and soon realized that Stub was really no match for him. Unable to escape his furious onslaughts, she was so badly battered that she died of internal haemorrhage the following day.

I was very distressed to receive this news, and as I sat on the verandah that night, gazing out onto the garden which was bathed in moonlight and took on an ethereal quality, my thoughts turned to a carefree little figure, only eighteen inches tall, galloping across the lawn with tail tucked well in and ears flattened back. When I closed my eyes, I could see it all so vividly, exactly as it had been five years ago. I couldn't believe that the effort taken to raise this little animal had been to no avail, and as the moon rose higher and a soft breeze stirred the leafy shadows, I pondered on the ease with which the vital thread of life can be snapped. Was it all so terrifyingly final? It comforted me to hope that somewhere in that great dark void, Stub would be given the chance to gallop again.

After the death of Stub, I hoped that I would be spared any more rhino orphans, which entailed the most work and the most problems and heartaches. But, this hope died abruptly only a few days later when Stroppy, a female calf, about nine months old, was found alone in the bush and was dumped on our doorstep. My heart sank further when I heard the inevitable bellowing from the carpark, and realized that I had another buffalo orphan as well; and the last straw was Punda, a minute zebra foal who had mistaken a striped minibus for its mother, and had been taken aboard by the tourists inside, and deposited at the Gate. To make matters worse, Stroppy the rhino fell desperately ill, and hovered on the brink of death for two weeks which entailed special nursing, and was extremely time consuming, for she even had to be lifted to her feet she was so ill. But my initial disappointment turned to pleasure when I got to know the three newcomers better, for all three were real characters.

Punda idolized Stroppy with a possessive and jealous passion which grew with every day that passed. He would fuss over

Stroppy, nibbling her horn, rubbing up against her, nuzzling her and incessantly watching her every movement, devoting every waking moment to the adoration of this unlikely companion. Stroppy received Punda's attentions with a bored indifference. Gradually, as she began to recover, her skin lost its dry, flaky appearance and took on a supple texture, and when she showed the inclination to play, we knew that our troubles were over.

A baby rhino playing must surely be one of the funniest sights to watch. Having a zebra and a buffalo involved in the game makes it funnier still. Such games were usually triggered off in the mud wallow, provided at the side of the house for Stroppy's benefit. Stroppy would roll from side to side with much leg kicking, snorting, huffing and puffing, leaping up every now and then to horn the ground before suddenly galloping off at top speed. Close on her heels would be Punda, resembling an Arab thoroughbred, nostrils flared and eyes wild, followed some distance behind by Boffin, the buffalo, lumbering along like an ungainly tank. All three would race round in a follow-my-leader, until Stroppy decided to turn on a sixpence, and would spin round without warning, coming to a dead halt, head held high and flanks heaving. Punda close behind, usually ended up by having to jump clean over her to avoid a head on collision, which never failed to take Stroppy completely by surprise and put her very much on the defensive. By this time, old Boffin would have arrived, just in time to collect the reprisals, and then, off they'd all go again, round and round, in and out, in a ridiculous merry-go-round that left the human spectators doubled up with laughter and having to dart for cover every time the cavalcade approached.

But, while all this provided good entertainment, it didn't do my garden much good, and so as soon as the three newcomers were past the infant stage, they were promoted to the main orphan herd. Strangely enough, Eleanor, who was usually not partial to rhinos, and had even disliked Stub, accepted Stroppy without any trouble, and seemed to quite like her. Her feelings towards Stroppy were reciprocated once the initial nervousness had passed, and Stroppy seemed to derive satisfaction from standing close to Eleanor. The games continue, but now with the elephants joining in as well, and Eleanor frantically rushing around trying to round up her unruly herd and establish law and order again! Punda's adoration of Stroppy has not waned, and although the same old problems are there in the future, for the moment everyone is happy.

Armageddon for Elephants

THE development of the Park had reached the stage when there were now over 1,000 miles of road and some 12 airfields to maintain. There were Lodges and more houses, entrance gates and ticket offices, new subordinate staff quarters and camping grounds, elephant-proof road signs at all intersections, boreholes, dams, bridges and causeways. There was a sophisticated radio network embracing all entrance gates and outposts as well as the mobile patrols. There were over 12 heavy earthmoving machines and 22 lorries and Land-Rovers, 25 stationary engines, generators, pumps, trailers and implements. Tsavo East had gone a long way since the days of 1 lorry and 6 labourers, and it had also outgrown its original headquarters. There was now an urgent need to modernize the entire complex and improve the workshop facilities in particular, for the task of the mechanical section was made even more difficult because of the extremely primitive conditions under which they were expected to tackle major repairs, and as the smooth running of any Park is dependent upon the proper upkeep of its transport and machinery, it was important that proper facilities be established as a matter of urgency.

Year after year, an appeal was made in the estimates for the provision of funds for this purpose, and also for the construction of a new and better equipped office block. David's own pokey little office now resembled a junk shop and housed a conglomeration of commodities that could be stored nowhere else, ranging from tusk butts to stuffed rodents, old files to Rangers' clothing. He could hardly move in it, let alone find anything, which resulted in the office work being neglected even more than before! But the only funds forthcoming as the result of these pleas was a sum of £700, which was barely sufficient for one office, let alone the rest.

There is a tendency for the headquarters of older Parks to snowball into sprawling villages as additional units are added to cope with the growth of the Park. David was determined that this fate should not befall Tsavo East, and when he designed the new headquarters, it was with an eye to an entirely new composite unit, provision being made for all the Park's needs. The sum of £700 was therefore not much good to him.

But once again fortune favoured Tsavo East from an unexpected quarter; the railways, who until now had been a thorn in our flesh, being responsible more often than not for the destructive fires that plagued us in the dry season. Perhaps it was to make amends for this that they offered David the Dak Bungalow on condition that he undertook to demolish the building and remove it completely from the station yard.

The Dak Bungalow was a Rest House, dating back to the turn of the century and the beginning of the railway, when no dining cars were attached to the trains. Many such places had been built at various stations along the line, where the train halted to enable passengers to leave their carriages for a leg stretch and refreshments. They could even spend what remained of the night there if they happened to be getting off at this point. The Dak Bungalow's reputation for excellent bacon and eggs and absolute cleanliness lingered even until my time, but in the early days it was a particularly popular meeting place. A loud bell was rung to warn of the imminent departure of the train, so that everyone could down their drinks hurriedly to scramble back aboard.

The Voi Dak Bungalow was a very substantial building with about ten good-sized rooms and teak doors shipped from India, for there had been no sawmills in those days. Massive steel girders supported an iron roof. David was jubilant, for with the material salvaged from the Dak Bungalow his new office block would become a reality, even with only £700. It seemed fitting also that this old building, steeped in history, which must have sheltered many distinguished personalities over the years, and seen many wild parties too, should not be entirely lost. Patterson, who hunted down the famous man-eaters of Tsavo, must have passed through those teak doors on many occasions. Our thoughts turned also to the man-eaters themselves, who had accounted for over 300 railway workers between them, and whose stuffed remains were reputed to be lying in an attic of a Chicago museum. We felt that they too deserved a place in Tsavo, and we hoped to try and bring about their return at some future date.

Today the remains of the Dak Bungalow are incorporated into Tsavo East's new elaborate headquarters. The rubble was fashioned into concrete blocks and formed the walls. Those same steel girders support the roof and the teak doors divide the rooms. Built on two levels and faced with the natural Tsavo stone, the headquarters comprises spacious offices for the Warden, Assistant Warden, Accountant and Clerk, a radio room and conference/operations room, an armoury, toilets and a guard-room with a built-in siren. The old offices, situated behind the new building, have been converted into much needed stores to house ivory and equipment. And in designing his own office, David indulged a long standing whim; he incorporated an aquarium into one wall, so that as he struggles with the always unpopular paperwork, he does at least get an illusion of coolness, which is more conducive to clear thinking!

There remains now the workshop to be brought up to date, and when this has been achieved, and the nerve centre of the Park has been modernized, there will be the satisfaction of knowing that this most important aspect of the development of the Park, its headquarters, has not been neglected but has kept pace with the progress in other spheres.

Almost ten years had now elapsed since the drought of 1960–61, and the intervening period of adequate rainfall had dissolved that tragic period into only a memory softened by the passage of time. The vegetation in those formerly devastated areas had made a spectacular and rapid recovery, and so had the rhino population. The soil that was once bare and exposed was now well-covered with a healthy tangle of perennial grasses, shrubs, legumes and *Acacia tortilis* seedlings, which had appeared to replace the commiphora, and which now in places stood close on ten feet tall. Since that time an air of well-being had pervaded. The rainfall had been well distributed; food had been plentiful, the flow of existing springs had increased and many others had appeared. The conversion by elephant of large tracts of former bush country into open plainslands with scattered trees was complete in many places south of the Galana River, and those animals that thrived best in a more open habitat had multiplied a hundredfold. Tsavo East was, in fact, now a far better Park, with a lot more to offer the tourist, than at any other period in its history. Even the Voi River, which used to be a purely seasonal stream, had taken to flowing for long periods, pouring continuously out of the dam over the spillways to wander through its lower reaches and end up in the sea. The main channel, however, had periodically silted up in a number of places, and the river had altered course on a number of occasions during that time, spilling over its banks in times of flood to snake a new channel elsewhere. This development affected one of the loveliest natural water-holes of the Park – Kanderi, a shallow depression by the Aruba Road which was normally filled by the first main floodwaters from the river. Kanderi was a paradise for waterbirds of every sort and was situated in a pleasing and peaceful setting, with the lush riverine vegetation behind, and the twin hills of Ndara and Sagalla mirrored in its glassy surface. Several twisted, typically Tsavo

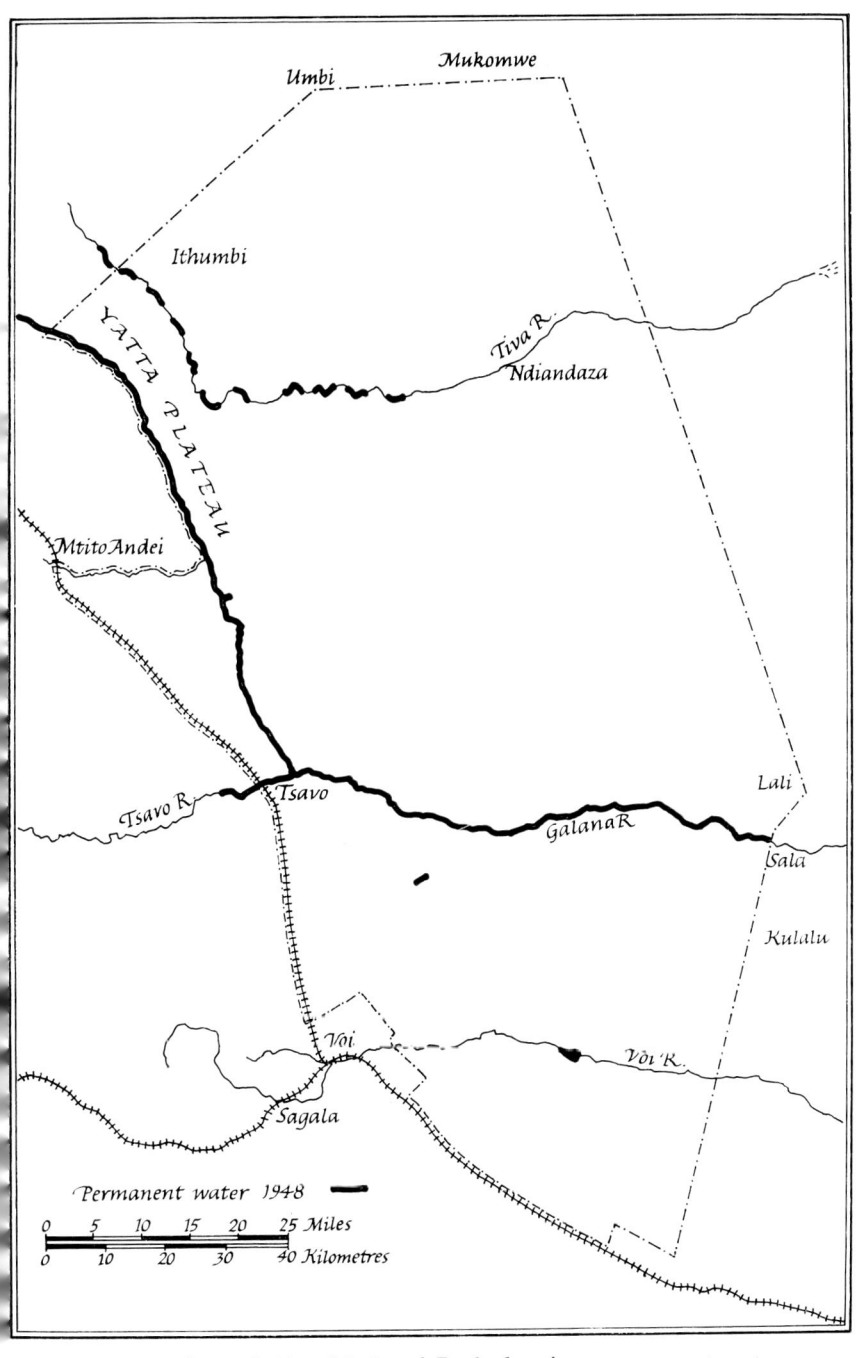

Map of TSAVO East National Park showing permanent water
in 1948

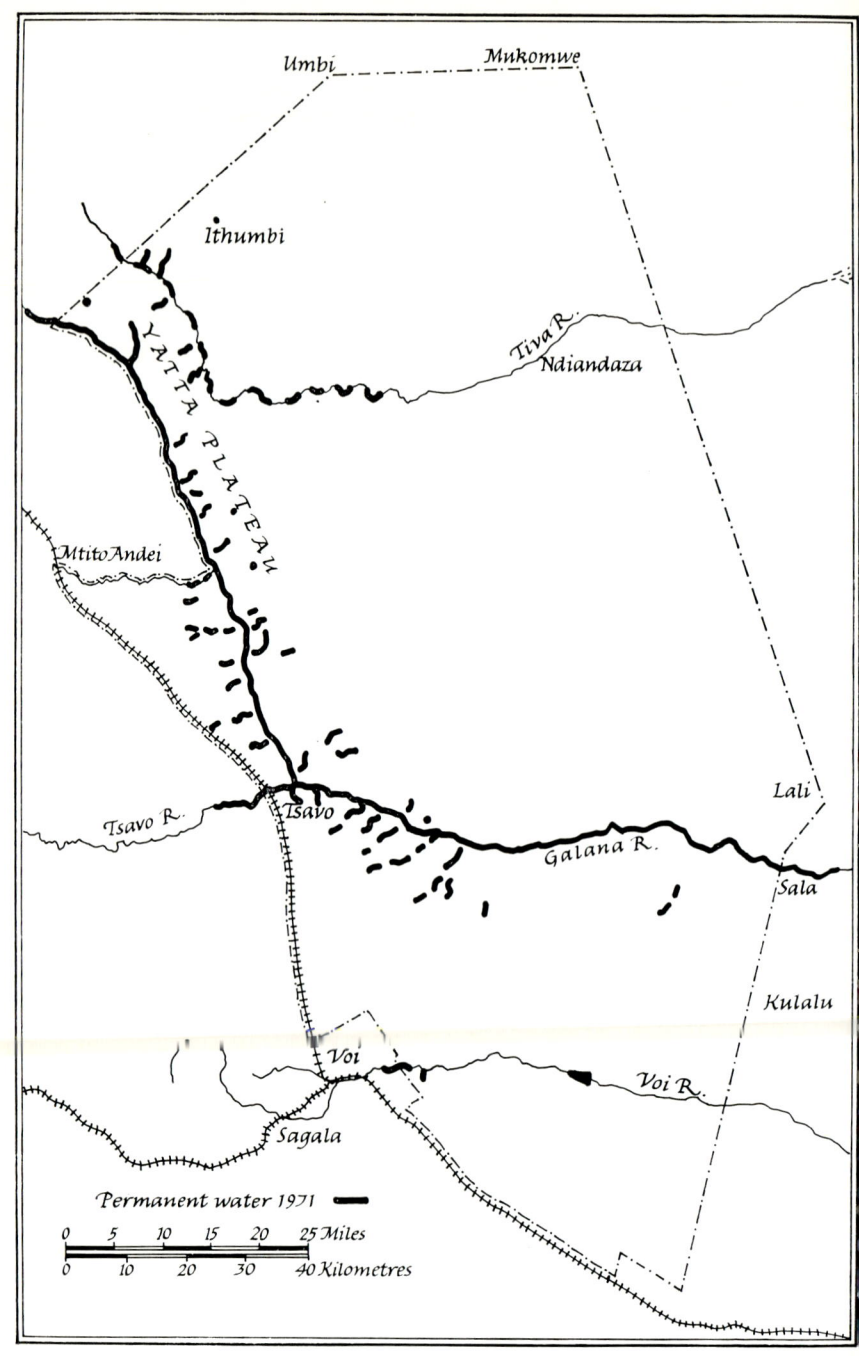

Map of TSAVO East National Park showing permanent water
in 1971

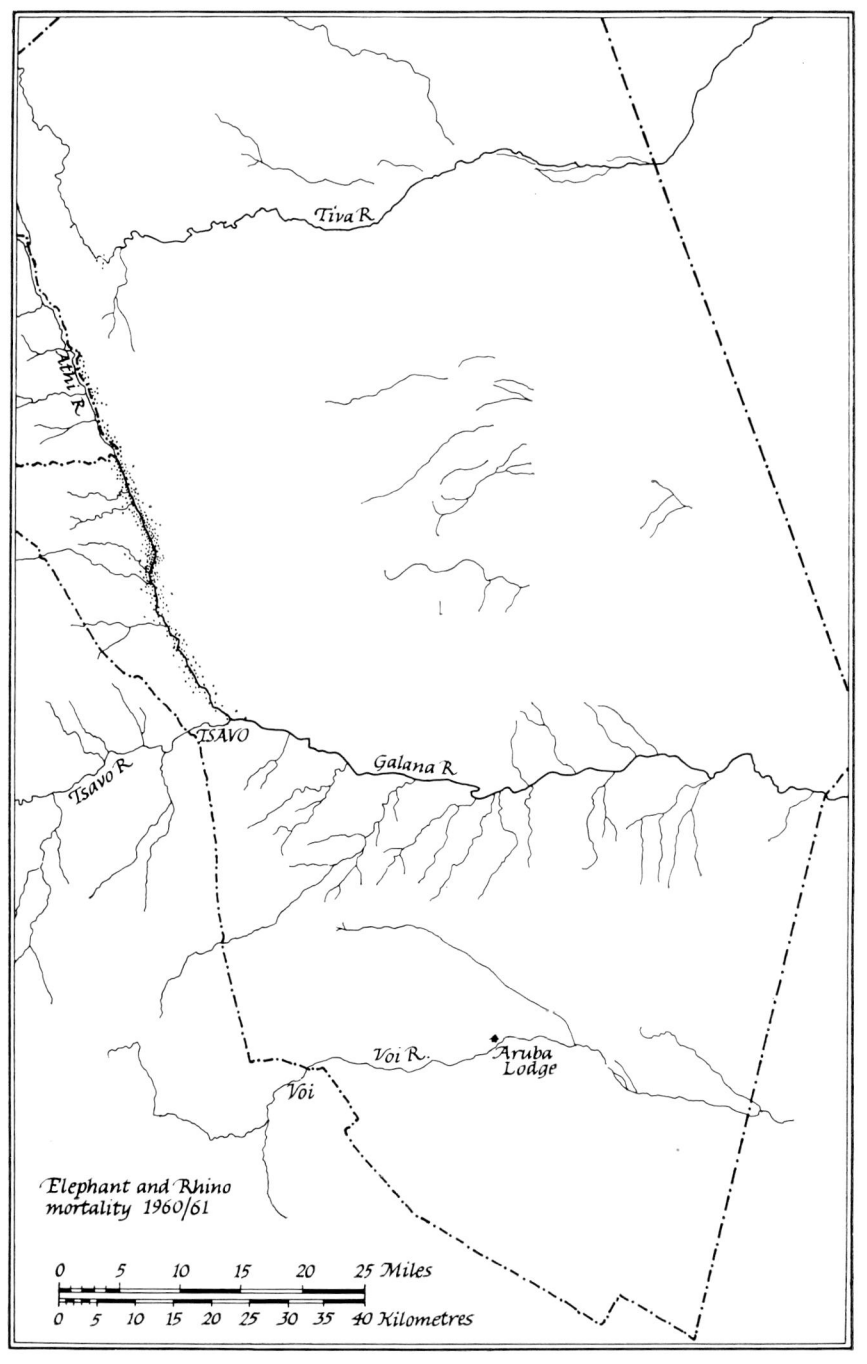

Map of TSAVO East National Park. The dots represent elephant
and rhino mortality in 1960/61

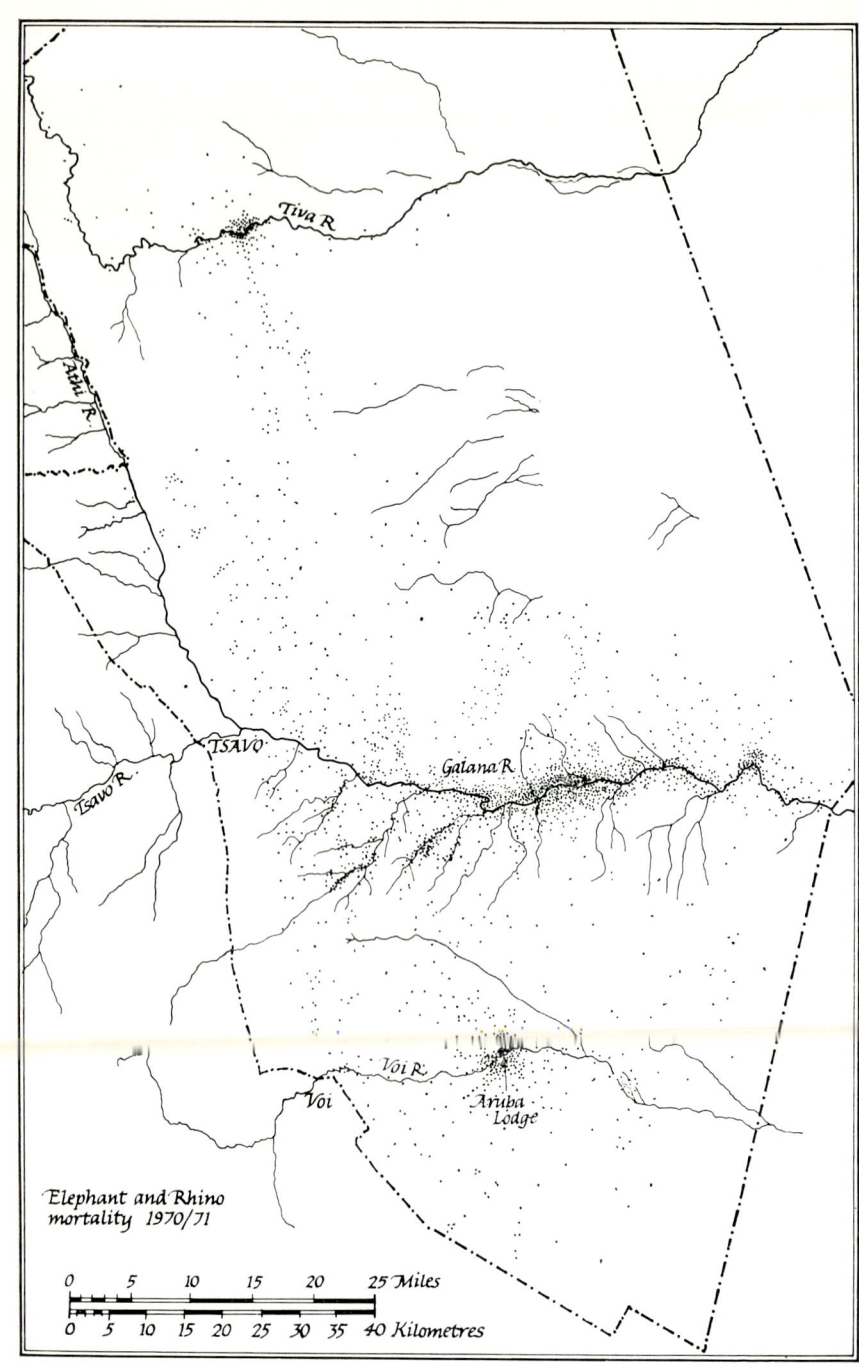

Map of TSAVO East National Park. The dots represent elephant
and rhino mortality in 1970/71

trees, projecting from the centre of the pool afforded a stark and picturesque perch for the snow-white egrets and the lovely sacred ibis, while wild duck and Egyptian geese, storks, darters and cranes gathered in the shallows to feed on the frogs and fish that were found there in plenty. Unfortunately, the Voi River took to flowing through this waterhole, and after a time Kanderi gradually began to silt up, until the reeds and grass broke the still surface and the waterhole became more like a shallow swamp. In 1969 the river reverted to its normal pattern and again ceased to flow during the height of the dry season, so that Kanderi began to dry up. A dry crust formed on the surface, but an accumulation of silt deposited beneath formed a treacherous bog of quicksands that proved to be a death trap for elephants.

Elephants are usually careful to avoid walking anywhere that may present difficulty, but for a time Kanderi was so deceptive that it fooled even them, and having broken through the crust, they sank down into the mud until they were hopelessly bogged and almost totally submerged, only a portion of the body and head visible above the level of the mud. Several elephants died in this way before we realized the danger now presented by Kanderi, for they were not easily visible from the road, but once we realized just how treacherous the waterhole now was, a close watch was kept on it and at least twelve other elephants were saved and extracted from the quicksands with the help of the large Michigan Bulldozer.

I am convinced that these animals understood that they were being assisted, for although they tried to struggle when they were first approached on foot, and attempted to flail the Rangers with their trunks, they all lay quietly to allow the steel cable from the tractor to be positioned around their bodies before being towed out bodily to firmer ground. Several of them, having struggled for many hours to the point of complete exhaustion, could not get to their feet unaided even then, and had to be lifted by the bulldozer blade. It requires quite a lot of effort for an elephant to stand up from a recumbent attitude, rising fore-legs first by throwing back the head, and many of the mud victims could not summon the strength for this initial effort. By digging the blade carefully beneath them to include a protective cushion of earth, it was possible to lift them to their feet with ease. Once up, not one of them made any attempt to attack either the tractor or the people standing about, but simply ambled slowly past everyone to disappear into the bush, as though they knew full well what had been done for them.

There was one small elephant, though, who didn't appreciate our efforts to extricate his mother and older brother from the mud, so that the rescue operations on this occasion were greatly hampered by repeated attempts to protect his family. Everyone was constantly being chased in all directions, until finally the calf had to be persuaded to leave the scene until the operation had been completed, and he could be reunited with his mother.

A similar hazard to Kanderi appeared at the top end of the Aruba dam when the waters began to recede there as well, and one rhino that was unable to get to its feet was transported back to headquarters cradled in the bucket of the tractor shovel, which looked as though it had been made to fit. Sadly this victim never did recover, but died shortly after arriving at its destination.

Rainfall figures can be extremely deceptive, and even misleading, for so much depends on how the rain actually falls. A succession of general showers at well-spaced intervals do far more good than one heavy downpour, and although the totals recorded may be identical, the results are certainly not. With well-spaced showers the vegetation has a long growing period, whereas a downpour will result in a good deal of wash and will bring on a green flush that will soon wither if there is no follow up. Again, in a vast area like Tsavo East, the rainfall is extremely variable. Some places can expect an average as high as twenty-one inches a year whereas others are lucky to record ten, and the way those ten inches fall is critical, and can mean the difference between life and death to hundreds of animals.

Two rainy seasons can be expected each year in Tsavo; the 'short' rains of November/December and the 'long' rains normally due in March, April and May. Here again, the terms 'short' and 'long' which differentiate the two pluvial seasons in East Africa, are misleading when applied in the context of Tsavo where the position is more often than not reversed. Here, it is the 'short' rains that are the longest and most reliable, while the 'long' rains, although usually the main rains for the rest of the country, are in Tsavo erratic and unpredictable, and if the November/December rains are poor, then there is cause for anxiety.

We got an inkling that we were due for another severe drought period at the end of 1969, when the rain, although recorded as only slightly below average in most areas, was nevertheless extremely patchy and poorly delivered. This was followed by very disappointing long rains during the first half of 1970 that were particularly poor in that portion of the Park lying east of a line drawn on the map from Voi to Lugard's Falls. Here less than two inches of rain was recorded during the first half of the year. Our fears that the ten year drought pattern was once more to be repeated were now confirmed, but what was extraordinary was the fact that on this occasion the indications were that the area that was likely to be worst hit, coincided with that which had escaped the effects of the 1960–61 drought, whereas the portion of the Park that had suffered such severe punishment at that time, was now in good heart and looked like escaping unscathed. It seems that everything in Nature usually evens out in the end and it was apparent that on this occasion conditions would be critical in the area lying east of Lugard's Falls. This included the Aruba area about which David felt especial concern, for he had been mainly responsible for its already desiccated appearance, when, following a fall of one inch of rain he had decided to burn the rank grass, anticipating further rain to bring up a fresh, green sward in its place. But not another drop had fallen since, so that the burn had turned out to be disastrous rather than beneficial. Never again, he vowed, would he be tempted into burning grasslands on the red laterite soil, which appeared to be far more fragile than the black cotton soil of areas of impeded drainage.

The term 'drought' when applied to events in Tsavo in

1970–71, is again misleading, for many people are led to suppose that a lack of water was responsible for the dire predicament in which the elephant and rhino found themselves in certain places. In some ways, this was true, but only insofar as a lack of water in the natural pans following the failure of the rains precluded the utilization by game of large areas of the Park which would normally have been available to them until the waterholes dried out. As it was, the game had to be totally dependent on the dry weather haunts for a much longer period than usual, and these areas had not had a chance to fully recover and rest from the previous season, the rains having been insufficient to promote any substantial regrowth in the vegetation. Food was therefore scarce, and became even more so as the dry season progressed and the pressure increased. The term 'famine' would probably afford a more apt description, for the Park was now better served for water than it had ever been before, and throughout the 1970–71 drought period the new springs and streams continued to flow, and in some cases even increased in volume. The term 'drought' on this occasion refers only to a lack of rain, which resulted not only in an absence of sufficient food, but an extremely low protein content in the parched vegetation.

The first to succumb in the stricken eastern belt towards the end of the 1970 dry season were elephant calves mainly from the 6 to 12 year age groups. It was distressing to see them in this tragic predicament. Emaciated and weakened, they lacked the strength to travel the long distances dictated by their elders, and finally refused to leave the permanent water at all, so that their mothers were forced to abandon them in order themselves to search for food further afield. Many fell prey to lions under these circumstances, but many others stood dejectedly beside the water until they died from malnutrition. This is not usually the long, agonizing end that people suppose, for Nature usually takes a hand in hastening the process. When an animal is low and in poor condition, the right climate prevails for a build up of the parasites in the blood, to which, under normal circumstances, the animal is immune, and either trypanosomiasis, one of the tick borne diseases, or pneumonia provide a quick release.

We have witnessed this fact on numerous occasions in animals rescued and brought back to headquarters, and which have appeared full of vigour one day, and dead the next. Stroppy was only one of many such cases, who would certainly have died had we not been there to administer a multitude of drugs.

About 300 elephant calves died before the short rains of November 1970 brought a reprieve, and it was interesting to note that those that were affected first were from the same age groups as recommended originally in the plan to reduce the population.

The abandoning of calves in times of stress was, of course, nothing new. It had occurred many times before in previous dry periods, and most of our elephant orphans had been acquired under such circumstances. Calves of this age are particularly vulnerable, having probably been recently weaned but not yet being as proficient at foraging for themselves as their elders. Handicapped by their size, they are often unable to make use of the available browse at a higher level, and they are the first to become weakened and unable to keep up with the herds. But, although such tragedies had been repeated many times before, more attention was now focused on this situation, for everyone was by now very conscious of the 'elephant problem' in view of all the controversy that had surrounded it over the years. Because of the severity of the drought we anticipated either a population crash or a mass exodus of elephants out of the Park, and although we were convinced that the steps taken by Nature to remedy the imbalance would, in the long term, be by far the most effective means of dealing with the problem, we suspected that it was bound to resurrect the controversy and generate quite a lot of heat in certain circles that were in favour of artificial manipulation, particularly amongst those who stood to gain in some way.

It is an odd quirk of human nature that whilst people are quite prepared to accept the organized death of large numbers of animals, easily satisfied by that magic word 'cropping', and easily persuaded of the need to avoid being emotional under such circumstances, they become extremely emotional when

Nature does the job, albeit efficiently, quietly, peacefully with no disturbance, and in a way that cannot be achieved artificially. There is suffering, of course, but suffering is a part of Nature, a process few creatures can escape and must endure at some time or another. Animals suffer every day in the wilds; they suffer when they are chased and killed by predators, when they are wounded in a fight, when they are sick, old, or handicapped and when they lose a loved one. But, they suffer too when they are cropped by being constantly harassed and terrified, for cropping, to be effective, must be continued once it has been embarked upon. Nature's way is selective and natural, and because of this offers a more permanent solution, removing the right numbers from the right age groups at the right time, which has a very marked effect on the future recruitment of the population. Furthermore, the stress to which the entire population is subjected triggers off adjustments in the breeding rate which cropping may even stimulate, aggravating the situation as Nature strives to recoup the unnatural losses brought about by unnatural means.

Although the short rains which started in November 1970 brought with them relief to the elephants in the eastern half of the Park, we suspected that this would only be temporary, for again the rains proved erratic and falls were well below average. By the end of the season the worst hit areas had recorded a total of only four inches for the entire year, which was one third of the normal average, and although there were the long rains still to come, which might interrupt the drought with a deluge, we feared that this was unlikely, and that we would have to steel ourselves to see through what was probably now inevitable. These fears proved correct, for the long rains were also inadequate to promote any noticeable regeneration of the vegetation in certain areas, with the result that deaths again started to occur in the same part of the Park as before, towards the end of August 1971, and escalated as the dry season dragged on. Again it was the calves that were first hit, but it was now not long before the cows suffered the same fate, when many of the old leaders succumbed. Very few bulls seemed to be affected, probably because they were more independent and unencumbered by feelings of responsibility for the young and the herd generally. The calves, weakened and sick, held up the cow herds, so that the cows suffered a rapid decline in condition as the result of malnutrition, and eventually were not able to summon the strength required to leave the permanent water in search for food. Most of the deaths occurred actually alongside or within reach of permanent water; along river banks and on game trails leading to water, which is only understandable, death from malnutrition being preferable to death from thirst. Day after day the merciless sun beat down from a brassy sky with a fierce, dry, intensity. It was pitiful to see the decline of such large numbers of elephant as they wearily hung around waiting for the end with mute resignation and silent apathy. It was not uncommon to see entire herds fast asleep beneath the scant shade afforded by a few gnarled trees, lying flat on their sides, or standing dejectedly in a huddle patiently waiting for an old, emaciated leader to make the move she plainly never would. It was pitiful also to see the attempts made to raise a dying comrade, or lift a far gone calf to its feet, but what was most tragic, day after day, was the great grief of a mother who had lost her calf, or a calf standing pathetically beside an inert mother. On one occasion a cow, whose small calf had collapsed, spent several hours painstakingly trying to urge it to its feet again. First, she pushed it gently with her back foot in the way that mother elephants rouse their sleeping offspring, and when there was no response, she tried supporting it between her trunk and foreleg. Alas to no avail, and when the calf had breathed its last, she felt every inch of the lifeless little body as though to imprint it on her mind forever, before turning deliberately away and slowly ambling off. One could sense the intense emotional suffering of this cow, which was far worse, I am sure, than the physical pain connected with the drought. Post-mortem examinations carried out on several carcases revealed that the stomachs were in fact full, but that the protein value of the contents was as low as two per cent.

Elephants are, by nature, gentle affectionate animals, whose ties towards one another are very strong, and who are capable of deep devotion and great loyalty. In some respects this was perhaps fortunate, for few individuals died alone, they were usually surrounded by friends when the end came. The rest of the herd would spend many hours standing beside a dying companion, and would not abandon her to her fate. Also many orphaned youngsters were probably adopted without any hesitation and cared for by foster mothers in the same way that Eleanor had accepted Bukaneza and Ndara extending to them all the devotion of a natural mother, even allowing them to suckle. So a bereaved infant was not necessarily doomed. Not so the unfortunate rhino baby, for the rhino too were very severely hit, suffering the same fate as had their brethren upstream in the 1960–61 drought. While an elephant baby will be accepted by the herd, a rhino calf is as good as dead itself once its mother has succumbed, and any strange rhino coming across a calf in this predicament, would be more inclined to kill it than adopt it. The plight of the poor rhino was therefore doubly tragic, for they were very much alone. And, as though all this were not enough, several luckless animals were found with snares around them to add to their misery, having obviously wandered into the neighbouring hunting blocks and tribal lands where anti-poaching measures had deteriorated to the point of non-existence. In fact, poaching outside the boundaries of the Park was now so serious that David, in his report, said, 'It must be recorded that poaching outside the boundaries has reached extremely serious proportions, and as the stocks of game outside the Park dwindle, there is a very grave likelihood that we must expect increased activity in the Park itself. It is essential that we be prepared for this onslaught and capable of containing it from the start. Poaching in the Park must not be allowed to get out of hand as it did in the 1950s, if the Park is to survive.' And although we had no jurisdiction over what went on beyond the Park boundary, David managed to obtain special dispensation for a combined Police/General Service Unit/Parks anti-poaching effort in the form of a series of raids, hoping that this might act as a deterrent. One elephant snared round the neck was dragging two enormous logs on the end of the wire cable which left deep ruts in the ground, and which had caused the wire to eat right through the flesh almost to the spinal column, until this unfortunate animal was put out of its agony. Likewise, a rhino was found wandering around with its head practically severed from its body by a wire snare. This despicable form of trapping wild animals inflicts untold suffering every year, and is, sadly, very much on the increase in Kenya.

Rapidly the drought toll mounted. It was particularly distressing for David, who could not escape being in the midst of it all, and who had either to visit or fly over the area every day to locate carcases and guide ground patrols in to recover the ivory and retrieve the lower jaws which were needed for ageing purposes. When you love elephant and understand them as he does, then you, yourself, participate in their suffering and can't avoid deep sorrow which, day after day, leads to depression and a feeling of despondency. This was not eased by the hordes of photographers, journalists and television companies that dogged his brief periods of relaxation for statements and sensational pictures, so that even then he could not escape thinking, talking and even dreaming about elephant.

But it was not only in the Park that the elephants died. The drought, which turned out to be the severest on record, affected many outside areas as well. The effects of it were felt over a very wide area to the north of the Park, where hundreds of head of cattle and even camels were dying from starvation. Elephant also died in the neighbouring Galana Ranch (where, incidentally, they had been cropped in Ian Parker's time), and even at Kolbio near the Kenya/Somalia border. In the Park itself, ironically, one of the worst hit areas was none other than Kowito, where the sample of 300 had been removed for Dr Laws, but even this had made not the slightest difference. It is therefore doubtful if the large scale die-off could have been averted, even if the elephants had been cropped as had been suggested. Due to the exceptionally dry conditions, there simply was not enough protein in the vegetation to enable an elephant to thrive, so that only the strongest could hope to last out until conditions improved with the onset of the rains, and any with impediments were ruthlessly weeded out. Nature selected the young, the old, the sick and the maimed, borne out by subsequent examination of the jaws of the drought victims, many of which carried deformities of some sort.

At this time, the survival or otherwise of the individual elephants in a herd depended to a very large extent on the leader. Those units led by an energetic and strong leader were in far better condition than those in charge of an old leader, who lingered too long at the water, and who felt the need to spend a lot of time resting. But, in the end, even the social structure of the elephant herds in the stricken belt collapsed completely. The family units disintegrated; cows took to wandering alone, abandoned and orphaned calves shuffled aimlessly about, making no attempt to join up with other elephants, but standing instead listlessly under trees. It became a struggle for individual survival, with only the most basic family ties withstanding the strain; that special relationship of a mother and her immediate dependent offspring. Very common also was a phenomenon peculiar to elephant, which is seldom seen under normal conditions, but which was a frequent practice amongst both calves and adults alike during the drought; their ability to extract quite large quantities of water from reserves in the stomach, by inserting the trunk into the back of the throat and sucking the liquid up into it so that it could be sprayed behind the ears and over the flanks, presumably as a means of regulating the body temperature. One young calf, standing disconsolately beside the body of her mother, was seen to spray herself in this way repeatedly from mid-morning until mid-afternoon when she was rescued and transported back to headquarters. Happily, this little elephant, called Sobo after the place where she was found, was one of those that had a happy ending, for out of Eleanor's large heart flowed comfort and love, so that Sobo now lives contentedly amongst plenty, a very different elephant from the emaciated, pathetic little creature that was brought in. Nor was she the only one we tried to help under like circumstances, but most of the others were too low to resist the shock of

274

capture, and succumbed within a few days. Eleanor always did her best, kneeling down to help us lift the drought victims, fussing over every new arrival, and adopting them all with no hesitation. Also, we never quite knew just how many elephants we would have each day, for she would collect odd waifs and strays down by the river, and bring them back home with her in the evenings. Sitting on the lawn in the evenings, we would be surprised to see five elephants filing up the hill instead of the usual four, and one day there were even six. Neither of the new-comers developed into permanent residents, however, being rather independent little bulls who opted out after several days, much to the relief of the attendant who had been given quite a rough passage by them. But then, one night, the lorry arrived with the minutest little elephant on board I think I have ever seen. The effort of bringing it into the world had cost its drought-stricken mother her life but the calf was strong and as soon as it had been lowered to the ground, shuffled from one person to the next ravenously hunting for a teat to latch on to. Although we realized that the chances of being able to rear this little elephant successfully were negligible, we decided to try nevertheless, this time with the help of Eleanor.

Eleanor was overjoyed at the newcomer and beside herself with excitement as it was pushed into the stockades. All her boundless tenderness and protective instincts welled up as she gathered the baby to her, rumbling endearments and feeling it all over with her trunk, coaxing it beneath her forelegs and encouraging it to suckle. I had meanwhile prepared a feed and, scrambling beneath Eleanor's tummy from the other side, substituted the bottle for Eleanor every time the calf attempted to suckle, which proved very successful.

The next morning the baby accompanied all the orphans, and its bottle of milk plus all the paraphernalia required for heating it, went along as well, for we had decided this time to try feeding the calf on demand. Every time it looked as though it wanted food, the attendant was instructed to dive under Eleanor and thrust the bottle into its mouth.

There was, however, one member of the orphan entourage

who hampered operations considerably – Bukaneza, who until now had enjoyed Eleanor's undivided attention and who didn't appreciate having his nose put out of joint. The main source of trouble was the presence of the bottle, which tantalized him beyond all measure, and drove him almost to distraction when he had to watch someone else enjoying what he felt should be for him. His frustrated yells rang out clearly throughout the day, so that it was possible to follow the orphans' progress at all times, and so obsessed was he with the bottle, he even refused to feed properly and began to lose weight.

But, as we had feared, sadly the little elephant also went on the decline, and every day it grew a little weaker, until the hollows in its cheeks and forehead took on that sunken appearance of the drought victims, and it looked hunched and sick. We knew that we had again failed, and that it was just a matter of time before the baby died, so we decided to put it down. The difficulty came in getting Eleanor to relinquish the calf. Every time we attempted to remove it, she became agitated and even annoyed, and simply refused to be parted from it. In the end the problem was solved by the arrival of another drought victim in a state of collapse, and while Eleanor was fussing around that, we were able to spirit away the little one and bring its suffering to an end. Fortunately, Eleanor was so concerned at the plight of the newcomer that she appeared not to miss the little elephant very much, or perhaps she realized what had happened. Another incident remains imprinted in my mind – an occasion when we were trying to revive a calf by getting it to take some milk and glucose, but were hampered by its inability to lift its head. Very deliberately, Eleanor took hold of one of its little tusks with her trunk, and lifted its head up for us, holding it in position for several minutes until I had poured the milk into its mouth. Was this really an accident, or did she understand what was needed?

Sadly, the only one out of many drought victims to survive was little Sobo, who rapidly became rounded and strong, stuffing lucerne into her mouth feverishly as though determined never to be hungry again. Bukaneza regained his privileged position as Eleanor's favourite once the little elephant had gone, and Sobo came a close second. She obviously had no intention of being orphaned twice, for although on one occasion she became separated from the others and was lost for a day, she arrived back at the entrance to the stockade the next morning and joyfully rushed in as soon as the door was opened. The reunion with the others after this brief absence and her obvious pleasure and relief were touching to see.

The one very interesting fact that emerged from the drought was that the elephant in the affected areas made no attempt to move to where conditions were less critical. We never quite knew until the time came just how the elephant would react under such circumstances; whether they would move out of the area *en masse* in search of better pastures, or whether their fixity to home ranges was such that they would remain where they were and die of starvation. As it turned out, the latter proved the case. Those elephant whose dry weather territory coincided with the minimum falls of rain during this time died where they were in large numbers, making no effort to search elsewhere. It appears therefore that in the dry season elephant are anchored to their dry weather haunts. This fact was illustrated very forcibly not only on the lower reaches of the Galana River and around the Aruba dam, but also in a very narrow corridor adjacent to part of the Tiva River in the northern area, which just happened to miss all the rain. Here many elephant died from malnutrition although had they moved only five miles either up or down stream they would have found a plentiful supply of food.

The fact that elephant appear to be so localized in the dry season is probably very fortunate, and again has perhaps been ordained by Nature to protect the environment from the depredations of a large, roving population that could eventually eat themselves out of existence, cleaning out one area and then systematically moving on to the next. As it is, the numbers must fall in line with the carrying capacity of their particular dry weather feeding grounds in a drought year, and when the rainfall in an area is very low, the elephant and rhino resident there in the dry season suffer. Theoretically, this periodic selective die-off not only ensures a healthy population, but should also allow the regeneration of the vegetation during years of normal rainfall to keep ahead of the demands upon it.

To date the only two species affected by the droughts in Tsavo have been the elephant and rhino, probably because these two particular species have attained their peak in numbers and now qualify for natural control. Certainly this is true of the elephant, but it could also be true in the case of rhino, for it was estimated that the Park contained well over 5,000 rhino, and although some 300 had succumbed west of Lugard's Falls during the 1960–61 drought, the population in that area was now healthy and in good shape while the position was being repeated, ten years later, east of Lugard's Falls. Quite apart from this, however, these two particular species are probably the most susceptible to drought condition due simply to their physiology. Whereas most other herbivores, many of which are ruminants, can extract the maximum benefit in terms of available protein from what they eat, the elephant and rhino have been endowed with a far less efficient digestive system, and as much as 6% protein has been recorded as having been passed in their droppings. Furthermore, due to their bulk, they also have to fall back on protein rich browse to supplement their diet in the dry season when the grass is tinder dry. And this, again, is probably no accident of Nature, for, after all, whereas other herbivores are subject to predation, what predator, apart from man, can curb the increase in numbers of these huge animals? Drought is probably Nature's only means of controlling such populations.

It would have been too optimistic to hope that Nature could work out her solution to the elephant over-population without an outcry. The Park remained open throughout this period, and as most of the roads are aligned to follow the permanent water, the public could see for themselves what was taking place. Newspapers carried headlines like 'The Vultures find a Mighty Feast', '1,000 Elephants feared Dead' and 'Drought kills Tsavo Elephants', and the reaction of most people was, understandably, one of shock, of dismay and of sadness. People began to say 'Why was this disaster allowed to happen?', 'Why was it not forestalled by shooting the elephants instead?' and 'What are the authorities doing about it?' And the sort of thoughts that crossed the minds of those with an eye towards personal gain was 'What a *waste*, both in terms of protein and hard cash.' The fact that nature was doing what the scientists had recommended should be done, but in a far more sophisticated and peaceful manner was eclipsed in a flood of emotional outbursts from the critics of the policy in Tsavo, who harked mainly on the 'wastage' angle and the fact that the Tsavo elephants could have been exploited rather than allowed to rot in the bush. The prospect of financial gain is, of course, always attractive and arguments that nourish this prospect can easily convince, but those who suggested this seemed to have lost sight of the very meaning of conservation, which, in a National Park especially, should be the foremost consideration. Conservation is defined thus by one of the top ecologists of our time: 'Maintenance of the energy flux is conservation; reduction of it is the opposite to conservation.' No one could, in all honesty, argue that the removal of large numbers of elephant from the habitat would be anything other than a reduction of the energy flux, and as such contrary to the fundamentals of conservation. The part an elephant plays in the ecosystem when it is dead is probably just as important as its living role in nature, and when one contemplates that it takes only five days for an unopened, entire elephant carcase to collapse, and only five weeks for it to disappear completely, leaving just an empty shell of dried skin and, of course, the bones as evidence; and this without the aid of scavengers and vultures, some idea of the magnitude of the energy flux is possible. Further illustration of the important role of insects can be gleaned from the work of Dr Malcolm Coe, who was seconded to Tsavo from Oxford University in order to study this aspect of the drought. He recorded no less than 84,700 insects in three kilos of elephant dung, and so, with this in mind, one can perhaps get a better appreciation of the astronomical numbers of organisms that must be active to complete the recycling of an animal like an elephant in such a short period of time. But, who can possibly begin to understand fully the complexities of this process and its impact on the environment? And when one considers that by the time the storm clouds gathered and the first drops of rain fell to alleviate the situation, over 5,000 elephants had been recycled in this way, and when one visualizes what has been returned to the soil by those 5,000 elephants, one will realize that the term 'wasted' would have been more appropriate had those elephants been removed from the Park, and converted into cash instead; always assuming, of course, that one has the well-being of the Park foremost in one's mind. Nothing was removed from Tsavo, and so how then can it have been wasted, and if it had been removed, how then could we have reimbursed the habitat?

We began to wonder if it would ever rain again when September dragged into October, and October crawled into November. Still the sun beat mercilessly down, baking the earth and casting a shimmering heat haze that tantalized with an illusion of water. The strain was beginning to tell on David, and the great responsibility hung very heavily on his shoulders. He had postponed his leave but in fact he needed it badly.

The first half of November came and went, and although the clouds banked up promisingly each afternoon and the humidity was oppressive, still the rain held off – and more elephant died. This pattern was repeated day after day until the end of November, when the heavens suddenly opened one afternoon and those elephant in the eastern belt that had had the tenacity to cling to life until this day, were spared. It seemed fitting too that this should have happened on David's birthday, for nothing could have given him greater pleasure, or raised his spirits so successfully as did that first shower of rain. Joyfully everyone hurried outside to savour the first life-giving drops; to watch them fall in a puff of dust on the powdery soil, and to see them come with ever increasing intensity until the ground began to glisten and the water creep along in little rivulets. One could almost sense the quickening of the earth;

the hidden excitement of the birds, the insects, the frogs and the animals and the revival of the poor old elephants, as that soothing liquid straight from heaven poured over their emaciated bodies and brought with it the promise of renewed life. And, before the week was out, not a single elephant remained in the stricken eastern belt. Somehow the survivors managed to trek again, and they deserted that area of so much misery to congregate where the heaviest falls of rain had made the *nyika* burst again to life. For although elephant may be very conservative in the dry season, as soon as heavy rain falls, even many miles away, they somehow seem to know and overnight they migrate *en masse*. How they know where to go will remain a secret of Nature, probably forever, but the fact that they do is demonstrated every single year in Tsavo. Water in the inland pans means that fresh feeding grounds are now accessible, and the elephant are able to fill their stomachs again.

It didn't take long for the drought sufferers to lose their gaunt appearance, their listlessness and apathy, and again to romp in rain-filled pools and plaster the sticky red mud of Tsavo on their bodies. The air was heady with the minty scent of freshly plucked lush vegetation in which the sap was rising, and this combined with the heavy stable smell of large numbers of elephant brought about an atmosphere of well-being and plenty that left one wondering what all the fuss had been about. But it was always like this in Tsavo: a land of fierce contrasts and extreme moods that are reflected in its animals, its vegetation and even its people. However, the mopping up operations after the drought entailed an immense amount of unpleasant work. There remained hundreds of jaws to be brought in, and hundreds of tusks to be recovered. The work of the scientists was only just beginning, but we felt that the most important aspect of ours had probably now ended. We were glad that we had been able to steer Tsavo East through this critical period in its history, and having weathered the storm, David felt an intense lethargy and a compelling need to get away from it all completely for a time at least. To escape elephant and ecosystems, commiphora and conversion cycles so that he could relax again and get things back into the correct perspective. We left the rain falling in Tsavo with an easy mind, and went as far from it as we could afford – to the southernmost tip of Africa. But, even there, gazing out to sea from Cape point, David said, 'I wonder how things are in Tsavo; whether the rains have been good; and whether enough elephant have died.' He paused and then added, 'I sincerely hope so.'

When we returned, we discovered that the short rains which started with such promise had not been quite so widespread as we would have liked, and had in fact again been below average. Would drought conditions be repeated in 1972 then? Was Nature not through with the elephants yet? The answers to these questions have yet to unfold, for no one can control the weather, but whatever happens, and however significant it may seem to us at the time, it will be but a flicker of Nature's eyelid, a fleeting moment in her vast design. Probably, the most effective way in which we can help is simply by keeping inter-ference to the very minimum, and assess and try to understand the trends with the words of Confucius in mind. 'Study the past if you would define the future.' The Galla graves, Lord Lugard's waterfall, Sir Frederick Jackson's 'open country' and the great numbers of eland seen by Selous, healed scars on ancient baobabs and swings in elephant populations revealed by old records; all are pointers to what we can expect, but one thing is important; that by carefully monitoring events in Tsavo, mankind will be able to see for the first time what happens to elephant populations that are left unmolested, even within 'unnatural' boundaries. For nowhere in Africa where a parallel situation has evolved has it been allowed to develop to its natural conclusion. Always man has intervened, and tried to manipulate events usually with disastrous results. Wherever elephant are represented in any numbers, there is a 'problem', and the combination of elephants and scientists produces perfect ingredients for a monumental problem! Would it not be pru-dent to set Tsavo aside as a natural 'control', so that we can take a lesson from Mother Nature in the management of animal populations, a lesson which would be of the utmost importance

to many wildife sanctuaries in Africa, and if the balance between species seesaws, as it probably will, and if changes occur,
let us not rush in where angels fear to tread, but keep the funda-
mental concept of conservation well in mind, as defined by
Dr Fraser Darling: maintenance of the energy flux. It is this
concept which has provided the guidelines for the management
in Tsavo East to date. Changes are bound to take place in any
biological community and some changes may, of course, be
triggered off as the result of compression of habitat through
human pressure outside man-made boundaries, but even so,
appreciating but a fraction of the marvels of which Nature is
capable, and the wonders she works to perpetuate each and
every species; and mindful also of our present limited understanding of the complex processes involved, any clumsy
attempts to redress the balance could be dangerous, well-
intentioned efforts towards this end serving only to perpetuate
an undesirable situation. There will be periods of drought, and
there will be floods; there will be anxieties, problems, difficulties
and doubts; and there must, unhappily, also be some suffering
before the correct balance of species is achieved in such circumstances. It remains David's contention that the conservation
policy for Tsavo should be directed towards the attainment of
a natural ecological climax, and that our participation towards
this aim should be restricted to such measures as the control of
fires, poaching and other forms of human interference that tend
to lessen the energy flux. It is his belief that herein lies the safest
course for the wise management of the Park, and indeed, in
a continent like Africa, for its very survival.

The International Convention of 1933 defined a National
Park as 'an area set aside for the propagation, protection and
preservation of wild animal life and wild vegetation, and for
the preservation of objects aesthetic, geological, prehistoric,
historical, archaeological or other scientific interest, for the
benefit, advantage and enjoyment of the general public.' However, the activities of 'the general public' are often inimical to
the best interests of the wildlife and vegetation in a National
Park and there remains the constant need for vigilance, so that

that which it was our intention to preserve is not degraded, defaced or even destroyed in our efforts to cater for the needs of the public.

A good deal of faith had been pinned on research in the early days, but it was becoming increasingly evident that far too much had, in fact, been expected of the scientists, and in very few instances had it been possible for them to contribute towards the management of the Park by providing the correct answers in time to such vital issues as the long term effects of fires, whether altering trends in the habitat were desirable or otherwise from the point of view of overall soil fertility, or even the impact such changes may have on populations as a whole of the different species, and so on. This was, of course, due in part to the fact that such problems required intensive research and study over a very long period before any sort of accurate picture emerged, with the result that the scientists invariably found themselves simply overtaken by events. This tended to lead to requests for more staff, more equipment and more facilities, until there was a danger that the scientists themselves would become yet another problem, even with the best intentions in the world. More and more motorcar tracks left the roads to lead to experimental plots, exclosures, inclosures, rain gauges and other scientific paraphernalia, or to facilitate the observation of animals being studied, with the result that the natural unspoilt beauty of the wild scene was further marred by the marks of man. David had come to believe that ideally the research effort in any National Park should be kept within reasonable limits, with a team comprised of dedicated 'naturalist' scientists who were prepared to forfeit burning professional ambitions in order to become part of the permanent staff of the Park's organization, content to confine their work to limited objectives of direct importance to the well-being of the Park as a whole. He felt that it was important that the scientists working in a Park should consider themselves an integral part of the general Park's effort rather than privileged birds of passage, and that they should strive to further only the best interest of the Park at all times. There was no doubt that in Tsavo one of the most important functions of research was to monitor care‑ fully all the changes as they occurred, so that information so gained could be applied elsewhere with the benefit of hindsight.

But, whatever the future held for Tsavo East's inhabitants, the behaviour of the elephants during the brief rainy period that followed the long dry spell differed somewhat from that recorded during more normal years. For, although it is usual for large numbers of elephant to congregate in areas where the heaviest falls of rain have resulted in a plentiful food supply, the pattern this took following the heavy mortality sustained by the population during the drought was rather out of the ordinary. Normally, concentrations of elephant in the greenest areas are rather loose; more an aggregation of many different family groups in a given area rather than a definite combination of those units into one entity, and although several units may join up temporarily to form larger herds, this tendency was more marked following the drought. The elephant amassed themselves into very large herds, which roamed the countryside in close unison, bowling over the few remaining commiphora trees as though beset by a compelling urge to complete the cycle they had embarked upon, despite the fact that food was, for the moment, abundant. Many of the trees appeared to have been felled for little apparent reason, and even Eleanor was seen to march purposefully up to a spindly commiphora struggling for a place on the hillside, and heave against it with all her might in an effort to dislodge it, although many of the smaller branches and leaves were well within trunk-reach had she particularly wished to eat them.

This tree-felling obsession inborn in all elephants is surely something that is ordained by Nature; a means of releasing the nutrients locked in the wood, which have been drawn up from the earth over the years, but there is no doubt that it is also triggered off by a simple desire to 'show off' during periods of heightened activity, and the bulls are by far the worst offenders in this respect.

By April and May 1972 there were many enormous herds consisting of 200 and more elephants to be seen in certain green belts of the Park. We wondered whether this could perhaps have been due to the fact that so many of the units had lost their old leaders during the drought, and also because the unit structure of the population had collapsed during that period of extreme hardship. Indeed, had one not known that some 5,000 elephants had recently died, one would probably have suspected that there had been an astonishing increase in numbers, so conspicuous were the elephants in these huge mobs. From the elevated situation of our house on the slopes of Mazinga Hill, we could watch the progress of some of the enormous herds in the Voi River valley below us, crawling like terracotta beetles across the green blanket of the low landscape, disappearing briefly into thickets to emerge again beyond.

There seemed to be a preponderance of bulls in evidence, and a definite sense of intense excitement among the herds. Very few adult bulls had died during the drought, and we now speculated as to how Nature would set about correcting any imbalance in the sexes. Certainly, serious fighting seemed to be a lot more prevalent than usual, and there were a number of monumental confrontations that had a fatal end for one of the contestants. There was a good deal of mating also, and chasing of reluctant cows. One was left with the impression that the heightened activity amongst the elephant population this year was something rather significant and out of the ordinary; part of what one scientist termed 'the most dramatic mammalian event ever to have been witnessed and recorded by man'.

What happens from now on to Tsavo's red elephants will be of the utmost importance and significance to conservationists throughout Africa.

Conclusion

I T seems unbelievable that nearly eight years ago I concluded another book by saying that as far as the Park was concerned, it was the end of the beginning, but for us, the beginning of the end. This seems to have more meaning on this occasion, but whatever the future holds for us, we are thankful that we have been able to guide the destiny of Tsavo up until this time. For David, his work has been his life, and I have been content to take second place, for it has brought him the peace of mind that comes with the satisfaction of achievement, and contentment with the knowledge that his labours have been for the cause closest to his heart. How true the words of Thomas Carlyle written over a hundred years ago: 'Blessed is he that has found his work. Let him ask no other blessedness.'

But, we have been blessed in other ways too that are probably worth more than all the material riches of life. We have been blessed with happiness, and blessed in living close to nature; blessed in being able to breathe pure unpolluted air and being

287

able to enjoy the unspoilt beauty of our surroundings. We have been blessed too in the companionship of our wild orphans, for through them we have been able to acquire a more balanced understanding of Nature. They have enriched our lives probably more than anything else.

But, there remains just one other prayer to complete the blessedness. A prayer that the Park will survive; that our successors will be able to carry on this work and that a lifetime's work will not have been wasted. A prayer that the animals this sanctuary harbours will be permitted to live out their lives in peace long after we have gone. 'If the Park is to survive . . .' How often has that been said! Our prayer is that it will. That it will somehow weather the turbulent times that threaten it in the future and endure to offer continued sanctuary to the remnants of those great wild herds.